MW01618178

Life of Fred®
Chemistry

Life of Fred®
Chemistry

Stanley F. Schmidt, Ph.D.

Polka Dot Publishing

ISBN: 978-1-937032-19-7

Printed and bound in the United States of America

Polka Dot Publishing Reno, Nevada

To order copies of books in the Life of Fred series,

visit our website PolkaDotPublishing.com

Questions or comments? Email the author at lifeoffred@yahoo.com

If you happen to spot an error that the author, the publisher, and the printer missed, please let us know with an email to: lifeoffred@yahoo.com

As a reward, we'll email back to you a list of all the corrections that readers have reported.

First printing

for Goodness' sake

or as J.S. Bach—who was
never noted for his plain
English—often expressed it:

Ad Majorem Dei Gloriam

(to the greater glory of God)

A Note Before We Begin

This is the high school chemistry course that I wish I had when I was first studying chemistry.

Several studies have shown that students who have had a high school chemistry course—like this book—do much better learning college-level chemistry.

The **TRADITIONAL WAY** to teach chem (and math and physics and biology) is have the student memorize thousands of facts. Some chem textbooks are like telephone books. One of them is actually 1200 pages long and costs more than $200. If you were teaching chem to a computer, this approach would be perfect. Computers can memorize (and retain) a million facts without breaking into a sweat.

The trouble is that many chemistry students are human. Opening a traditional chem textbook at random I found, "Gallium and In occur only in traces in Al and Zn ores. Thallium, also a rare element, is recovered from flue ducts from the roasting of pyrite and other sulfide ores." That's nice, but hundreds and hundreds of pages of fact, fact do not add any meaning to your life.

And even if you do memorize mountains of chem facts, you will forget 90% of them a month after you finish the course.

Life of Fred: Chemistry requires no sitting down and memorizing. You will learn that H stands for hydrogen and Na for sodium because you will be using those abbreviations often enough. All of the exercises are open book.

WHAT YOU WILL NEED

A scientific calculator that has keys marked sin, cos, log, and ln. They cost less than $20 and will also be used in advanced algebra, trig, and calculus. This is *not* the super expen$ive graphing calculator.

Enough of beginning algebra that you know that 10^{-3} means $\frac{1}{10^3}$ and can solve $9.444 = 3.222x + 3$, and know that $(4^4)(4^5)$ does *not* equal 4^{20}.

Conversion factors were taught in:

Life of Fred: Fractions
Life of Fred: Decimals and Percents
Life of Fred: Pre-Algebra 0 with Physics (on 28 pages)
Life of Fred: Pre-Algebra 1 with Biology
Life of Fred: Pre-Algebra 2 with Economics

Chemistry is where conversion factors are used a lot. It will be taught as if you had never seen it in the five previous books.

MY APPROACH TO TEACHING CHEMISTRY

In order to be a well-educated adult, you probably don't need to learn what a racemic mixture means.* But knowing the really basic stuff—such as that the chemical symbol for carbon is C and that organic compounds are those that contain carbon—is common knowledge. It's like knowing that World War I came before World War II.

A big part of high school chemistry is in learning a scientific attitude—how to approach new problems. Learning all the tricks and rules for doing chem problems is much less important than understanding how chemists could discover the weight of a single helium atom. In the beginning of this book Fred is going to find that weight *using only ordinary scales.* He will find that a helium atom weighs approximately 0.00000000000000000000000666 grams.

HOW CHEM IS DIFFERENT THAN MATH

You have to call 911 more often. Shattering test tubes, burns, and explosions all add to the excitement.

HOW CHEM IS LIKE MATH

If it is taught right, it can be a lot of fun.

Welcome to the adventure!
Stan

---

* (ray-SEE-mick) An organic compound that has equal amounts of dextrorotatory and levorotatory molecules = a racemic mixture. Discussions of racemic mixtures rarely come up on prom night.

Contents

Chapter One
Chem Lab Down the Hall

Fred taught his math classes in the Archimedes Building. It was the only building on the KITTENS University campus with an auditorium classroom large enough to hold the 805 students for his beginning algebra class.

Fred was explaining exponents.

He had written on the board: $\frac{x^m}{x^n} = x^{m-n}$

He asked, "What if m equals n?"

He wrote $\frac{x^n}{x^n} = x^{n-n}$

"And this simplifies to $1 = x^0$. Anything to the zero power is equal to one."

Fred was about to show that if you let m equal zero, $\frac{x^m}{x^n} = x^{m-n}$ would turn into $\frac{1}{x^n} = x^{-n}$, when Joe staggered into the classroom.

He was bleeding.

A broken piece of glass tubing was stuck in his wrist. Three algebra students in the front row passed out. They were obviously not pre-med students.

An algebra student, who was a veteran with overseas battle experience, stood up and said, "Let's see what you got here, sonny. Looks like a bit of a flesh wound. Nothing serious."

Fred passed out.*

The veteran walked Joe down the hall and back to the chem lab where Joe had been working. He located the first-aid box and dressed the wound.

---

* The last part of Shakespeare's *Hamlet* was very much like this scene in *Life of Fred: Chemistry*—bodies all over the place and lots of blood.

On the chem lab wall were giant posters. Joe hadn't noticed them.

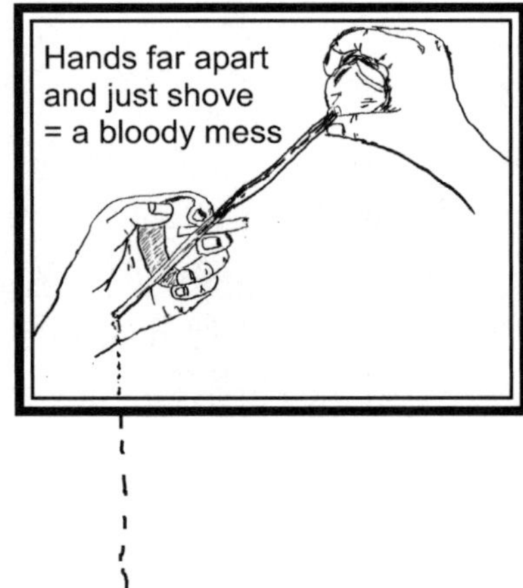

Some student had "decorated" the poster on the right.

Fred and his algebra students followed the trail of blood to the chem lab. He was still feeling a bit wobbly, but he was concerned about Joe.

Fred asked Joe, "Where's the rest of the chemistry class? You weren't working alone here, were you? That's not permitted for beginning chem students. And where are your shoes?"

That was too many questions for Joe to answer. Often when asked a bunch of questions, people just answer the last one. Joe said, "I'm barefoot."

"I know that," Fred said. "It's a standard rule in chem labs: NO BARE FEET. There might be broken glass on the floor."

Joe explained that he had arrived at the lab a half hour before it was scheduled to begin so that he could have extra time to mess around with the equipment.

There was broken glass all around the places where Joe had been working.

The veteran told Joe, "Hey man. Looks like you're set up for a second Purple Heart."

Fred translated, "Joe, your feet. . . . I mean. . . . bleeding." Fred passed out again. Teaching mathematics is one of the most accident-free professions on earth. Fred wasn't used to seeing any kind of trauma.

The vet picked up Joe and set him in a chair. Another student swept up the glass. One student had called KITTENS Hospital telling them that there was a medical emergency in the chem lab in the Archimedes Building.

The medical team was used to visiting the chem lab and seeing wounds from broken glass, chemical burns from concentrated sulfuric and nitric acids, and inhalation of toxic fumes that contain chlorine (Cl), bromine (Br) or hydrogen sulfide (H_2S).

When they arrived, they saw Fred passed out on the floor and guessed that he had breathed toxic fumes. They put his 37-pound body on a stretcher. He awoke and hopped off the stretcher and said, "I'm okay. I just fainted. It happens a lot." (It does happen more frequently to people who haven't eaten in *days*.)

They left.

Joe picked pieces of glass out of his feet and tossed them on the floor. The student who had been sweeping kept sweeping. Everyone who knew Joe, knew that he might step on that same piece of glass a second time if it wasn't swept up.

The veteran bandaged up Joe's feet and told him he should "see a medic pretty soon."

Fred knew Joe didn't understand army talk and translated, "Joe, you should see a doctor today."

Joe said, "I thought a medic was a doctor for a headic [headache]." He popped a couple of jelly beans into his mouth and chewed them.

The veteran pointed to another poster on the wall.

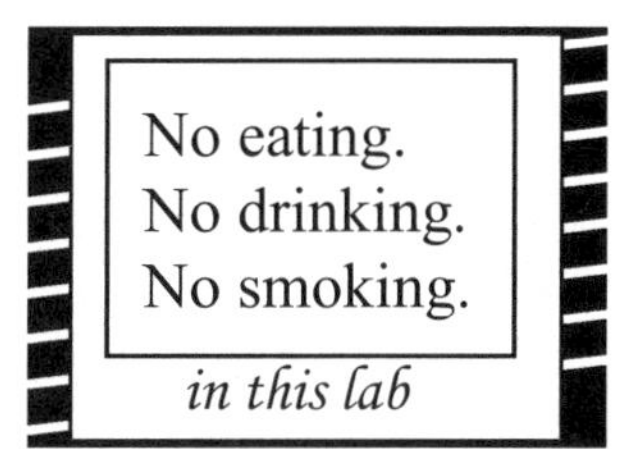

Joe protested, "But I'm not fat. That sign is for people who eat too much."

The algebra students couldn't believe what they were hearing. They left the lab. It was getting near ten o'clock when the beginning algebra class was scheduled to end.

Fred and Joe were alone. Fred explained to Joe that there were poisonous chemicals in the lab and that getting even little amounts of them in his body—Fred didn't use the word *ingesting*—would be bad for him.

Fred pointed to a bottle of sulfuric acid (H_2SO_4) and sang the famous chem song:

Johnny was a student. ♪
Johnny is no more.
What Johnny thought was H_2O ♬
Was H_2SO_4.

Joe didn't understand the song, but he told Fred that he knew why you shouldn't smoke in a chem lab: It gives you lung cancer.

Fred shook his head. The truth was that even if you smoke a lot in a chem lab, there is a good chance you will never get lung cancer! Instead, you will be burned to death or blown to bits. Organic compounds (those that contain carbon), such as methane or butane, are highly flammable. Fumes of some solvents can be explosive.

The students for the ten o'clock chem lab started to come in. Today there were supposed to work with glass tubing. The two exercises were to learn to insert glass tubing into black rubber stoppers and to practice heating the tubing over a Bunsen burner flame and bending it into a right angle.

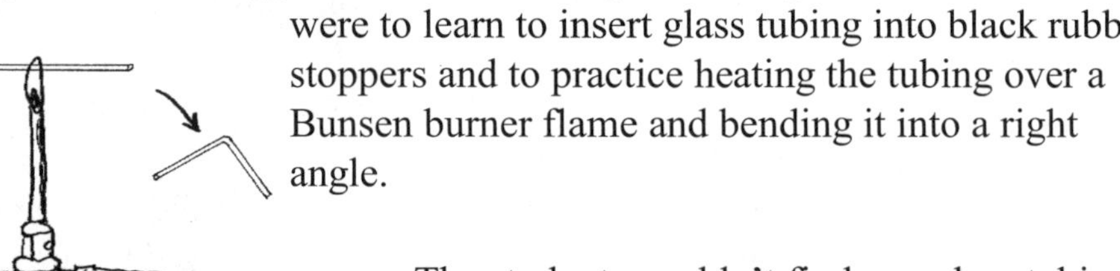
Bunsen burner

The students couldn't find any glass tubing. Someone must have used it all up. (We are not going to name any names.)

Three of Joe's fingers were badly burned. The experiment manual had clearly warned: BE CAREFUL YOU DON'T TOUCH THE MELTED GLASS. As an adult, Joe still uses kids' scissors. They don't have pointy ends.

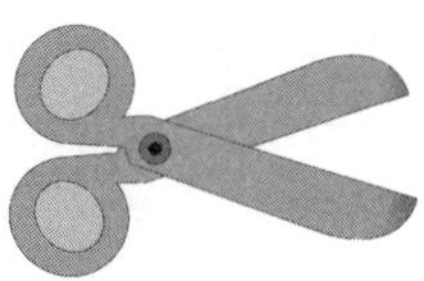
scissors for Joe

Please take out a piece of paper and answer these questions in writing before looking at my answers on the next page. It is much easier just to look at the questions and turn the page and look at the answers—but you won't learn half as much if you do that.

Your Turn to Play

1. Express as a decimal 10^{-2}.
2. Express as a fraction in simplest form 2^{-4}.
3. Simplify 98396^0.
4. Without looking back in the text, list as many of the chem lab safety rules as you can.
5. Can you guess why the veteran told Joe that he should see a doctor soon?

For fun, here is a list of the elements we have encountered so far.

H	hydrogen	
He	helium →→	Used to blow up balloons that float in the air.
C	carbon →→	Every organic compound contains carbon.
O	oxygen →→	About 21% of the air you breathe is oxygen.
Cl	chlorine	
Br	bromine	
S	sulfur	

There are only 90 naturally occurring elements. Every particle of matter is made up from these 90. You have already met seven of them. Learn about dozen more and you will know all the important ones. You will probably never need to know about thulium (Tm).

.......COMPLETE SOLUTIONS.......

1. $10^{-2} = \dfrac{1}{10^2}$ since $x^{-n} = \dfrac{1}{x^n}$

$= \dfrac{1}{100} = 0.01$

2. $2^{-4} = \dfrac{1}{2^4} = \dfrac{1}{16}$

3. 98396^0 equals 1 since anything to the zero power equals one. $x^0 = 1$

4. A. Read any safety posters or warnings in experiment manuals.

B. When inserting glass tubing into rubber stoppers, lube the tube with water, keep your hands close to each other, and use a twisting motion.

C. No bare feet.

D. Sweep up any glass on the floor.

E. Work under supervision—not alone.

F. Avoid chemical burns and toxic fumes.

G. No eating, drinking, or smoking in the lab.

H. Don't touch melted glass.

5. Joe should see a doctor: (1) to remove any glass splinters that were still in his feet, (2) to clean the wounds and perhaps apply antiseptic to prevent infection, and (3) perhaps to stitch up the larger lacerations (fancy word for jagged tears or wounds) so that they will heal better, and (4) to treat the burns on his fingers.

Chapter Two
Way to Go!

Fred had to get back to his auditorium classroom for his ten o'clock advanced algebra class. Seven hundred eighty-four students would be waiting for him. The beginning algebra students told the advanced algebra students what had happened to Joe.

Fred's Teaching Schedule
8–9 Arithmetic
9–10 Beginning Algebra
10–11 Advanced Algebra
11–noon Geometry
noon–1 Trigonometry
1–2 Calculus
2–3 Statistics
3–3:05 Break
3:05–4 Linear Algebra
4–5 Seminar in Biology, Economics, Physics, Set Theory, Topology, and Metamathematics.

They all flocked down the hallway, following the trail of blood that Joe had left, and stood at the doorway to the chem lab. Fred couldn't get out.

Joe pointed to the bandages on his wrist and on his feet. He liked to be the center of attention. He was drinking a can of Sluice to wash down the three sugar doughnuts he had just consumed.

One student looked at Joe and yelled, "Way to go!" Recently, Fred had looked up the word *go* in his dictionary.* There were 84 different definitions. The 27th definition was *to die.*

The chemistry teacher, Bob Bunsen,** and his 25 students were having a little trouble concentrating with Fred in the room and 784 advanced algebra students at the door. There were 25 Bunsen burners on the lab tables and 25 black rubber stoppers. And all the students were complaining that there wasn't any glass tubing.

Robert Bunsen

Bob didn't know what to do. The first thing he did was to tell Joe to stop eating and drinking in the

---

* *Random House Webster's College Dictionary*

** Bob Bunsen's great-great grandfather was Robert Bunsen, the man who invented the Bunsen burner. In the 1860s Bunsen also discovered two of the 90 naturally occurring elements (caesium and rubidium).

chem lab. He said, "Joe, eating is a no-no." No one used the word *deleterious** when talking to Joe.

Bob noticed that the garbage can was half-filled with broken glass tubing. He knew that Joe hadn't put it there because he never cleaned up after himself. Bob was having a very bad day—not harmful, not deleterious, just bad.

He turned to Fred and said, "I've got a swell idea."

Fred blinked. The word *swell* was so outdated. It was popular years before *cool*. And *cool* has been superseded by a half dozen other words.

Bob said, "There's not enough room here in my chem lab, and, besides, we are out of glass tubing. Let's all go down to your auditorium classroom and do our teaching there."

"But there's no chemistry equipment in my classroom," Fred said.

"I can fix that. One of the lab tables is on wheels. We can push it down the hallway." Bob loaded that table with all the standard chem equipment and three students shoved it through the doorway parting the Red Sea of algebra students.

In five minutes Fred was standing behind a large lab table. The audience was quiet, waiting to see what Fred would do. Fred looked around and wondered **Where's Bob?** Fred spotted him in the back row waving at him. Bob was going to have a delightful hour watching Fred do the teaching.

Joe was sitting in the front row opening a fresh bag of jelly beans. The cellophane bag ripped, and jelly beans scattered all over the floor.

Fred had read several dozen books on chemistry but was uncertain where to begin his lecture. He decided to start at the beginning.

"Atoms . . ."

Joe's hand shot up. "I gotta question."

---

* Diction means using the best word for the occasion. We tell Joe that something is *bad*. That's a nice short word indicating that there is something negative about it. If we use a longer word, *harmful*, we indicate that there are some bad consequences involved. Having a bad anger problem can be harmful to relationships.

A precise word to described what might happen if you eat in a chem lab is *deleterious*. That means harmful to health.

"Yes. What is it?"

Joe tried to remember what his question was. He began, "I was thinking . . . I mean . . . you know . . . everyone says that everything is made up of those little things, those atoms. I've never seen one. *Do you believe in atoms?*"

The class was absolutely quiet. Bob was so grateful that he didn't have to answer that question.

Fred said, "Yes."

Joe's second question was the tough one, "Why?"

This wasn't a chemistry question. This wasn't a math question. This was a question in epistemology.

Time Out!

Philosophy is traditionally divided into five branches: metaphysics, epistemology, ethics, politics, and esthetics.

Metaphysics: What really exists?

Epistemology: How do I know it exists?

Ethics: What then should I do?

Politics: What should I do in society?

Esthetics: How can I take what I know about reality and recreate it in the minds of my observers using literature, painting, music, drama, dance. . . .

Fred decided *not* to give this "big answer": My epistemology, like that of most scientists, requires that my conclusions regarding questions such as the atomic structure of matter are grounded in the formulation of models that are most closely aligned with the repeated observations of individuals whose metaphysics and ethics are in my estimation most congruent with mine.

Fred, instead, told Joe, "Atoms best fit what we've observed."

Joe frowned. Even this was too complicated for him. "But I've never seen an atom—even with a microscope."

Fred said, "Neither have I, but what scientists have observed does point to atoms. I'll ask you, Joe, a question. Do you believe you have a stomach?"

"Yeah. Sure. That's where my jelly beans go."

"But have you ever seen it?

"Well . . . no."

"Do you believe that Japan exists?"

"Of course. I have seen Japanese cars."

"But have you ever seen Japan with your eyes?"

"Well . . . no. But I know it's there even if I haven't seen it."

Fred smiled. "That is my point exactly."

Now Fred could continue his lecture. "Atoms are really small."

Before Fred could say another word of his lecture, Joe held up a ruler.

"Good," said Fred. "Do you have a mechanical pencil?"

Joe searched his pockets and couldn't find one. A student next to him handed Joe one. Joe hadn't been taking notes.

"Now draw a line."

"Please use the ruler."

"What is the size of the pencil lead?"

Joe didn't know. The student next to Joe said, "It's a 0.5 millimeter lead."

Fred said, "Now if you look on your ruler there are two scales. One scale is in inches. That's the one you use to measure the fish you catch."

"Yup." Everyone knows (who has read other *Life of Fred* books) that Joe likes to go fishing.

"The other scale is the metric scale. It's the one that almost every country uses, except for the United States, Liberia, and Myanmar. The metric system is so much easier to use because everything works in powers of ten. Scientists, even in the United States, use metric measurements."

Joe didn't look very happy. "But . . . what do you call the regular old measurements in inches, feet, yards, and miles?"

"It's called the English system, and before you ask, I'll admit that in England they don't use the English system anymore. They've switched to the metric system. They got tired of trying to add $2\frac{5}{8}$ inches plus $6\frac{3}{4}$ inches. They got tired of trying to convert $3\frac{1}{6}$ miles into yards."

Please write out your answers first before you look at my answers on the next page.

Your Turn to Play

In this course, you are allowed to use a calculator if you want to.

1. Which is easier to calculate? A) $2\frac{5}{8}$ inches + $6\frac{3}{4}$ inches
   B) 3.07 cm + 5.8 cm
2. Take your pick. Do either A) or B) from question 1.
3. H_2O is the formula for water. Is water one of the 90 naturally occurring elements?
4. Is helium (chemical symbol He) one of the elements?
5. Express as a decimal 10^{-6}.
6. Which is larger: 3^{-2} or 2^{-3}?
7. A meter is a little longer than a yard (about 39 inches). Six meters would be abbreviated as 6 m.

   A centimeter (cm) is one-hundredth of a meter.

   Find the values of x and y:

   1 m = x cm

   1 cm = 10^{-y} m

.......COMPLETE SOLUTIONS.......

1. It's my opinion that B) 3.07 cm + 5.8 cm is a lot easier to compute than A) $2\frac{5}{8}$ inches + $6\frac{3}{4}$ inches.

2. I will do both of them.

A) $2\frac{5}{8}$ inches + $6\frac{3}{4}$ inches

If we are going to add fractions we need to change to a common denominator.
The smallest number that 8 and 4 evenly divide into is 8.
That's the least common denominator.

$$= 2\frac{5}{8} + 6\frac{6}{8}$$

$$= 8\frac{11}{8} = 8 + 1 + \frac{3}{8} = 9\frac{3}{8}$$

B) 3.07 cm + 5.8 cm = 8.87 cm (Did you need a calculator for this?)

3. H_2O consists of two atoms of hydrogen and one atom of oxygen. Because it consists of two different atoms, it can't be an element.

Some time in this book, we will get around to showing how to break water down into two gases: hydrogen and oxygen. When two (or more) atoms combine together to make a **molecule**, the resulting compound is usually quite different than the atoms that made it up.

4. Helium (He) consists of a single atom. Helium is an element.

5. $10^{-6} = \frac{1}{10^6}$ = one millionth = 0.000001

1/10 = 0.1
1/100 = 0.01
1/1000 = 0.001
1/10,000 = 0.0001
1/100,000 = 0.00001
1/1,000,000 = 0.000001

6. $3^{-2} = \frac{1}{9}$ $2^{-3} = \frac{1}{8}$ To determine which fraction is larger, one method is to *pretend* you are going to add them.

$\frac{1}{9} = \frac{8}{72}$ $\frac{1}{8} = \frac{9}{72}$ So $\frac{1}{8}$ is larger.

7. 1 m = 100 cm

1 cm is one-hundredth of a meter, which is 10^{-2} m.

Chapter Three
Gold Atoms

Fred took Joe's sheet of paper and held it up for the class to see. He had to fold the paper to hide the food doodling that Joe had drawn with a crayon on the top of his paper.

"I'm going to show you how big an atom is," Fred said.

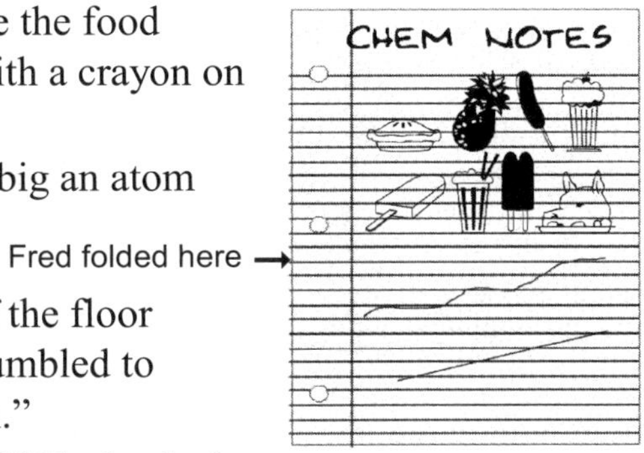

Joe picked up a jelly bean off the floor and popped it into his mouth. He mumbled to himself, "I gotta keep up my strength."

"Which atom would you like?" Fred asked. "You already know hydrogen (H), helium (He), carbon (C), oxygen (O), chlorine (Cl), bromine (Br), and sulfur (S)."

Joe wasn't sure which one to pick. "Is peanut butter an element?"

"No."

"Is gold an element?"

"Yes. Gold (Au) is an element. The chemical symbol Au comes from the Latin word for gold: *aurum*. I'm going to compute how many gold atoms lined up in a row would equal the width of that 0.5 millimeter pencil line."

"Are we allowed to guess before you figure it out?" Joe asked.

"Sure."

"I'm going to guess that it'd take a hundred big old fat gold atoms all lined up in a row to equal the width of my skinny pencil line."

Fred shook his head. "Atoms are really small."

Joe guessed again. "A thousand?"

Fred shook his head.

"A zillion."

Fred explained that a zillion is not a number.

He opened the chem textbook that Bob Bunsen had left on the lab table. In the appendix he found, "Au—atomic radius = 1.44 Å."

Fred had a lot of explaining to do.

"First of all, Å is the abbreviation for **angstrom**. One Å is equal to 10^{-10} m. That's one ten-billionth of a meter. It's much nicer to say that the radius of an atom of gold is 1.44 Å, than to write that the radius is equal to 0.000000000144 meters. If chemists had to write all those zeros all day long, they would never get any work done.

"An atom is roughly the shape of a sphere (a ball). The radius is the distance from the center to the surface. The width of the atom would be the diameter of the sphere. The diameter of a gold atom = 2.88 Å. (The math: 2 × 1.44 = 2.88)

"So we have to find out how many 2.88 Å would equal 0.5 millimeters."

Joe raised his hand. "My pencil doesn't say 0.5 millimeters. It says 0.5 mm."

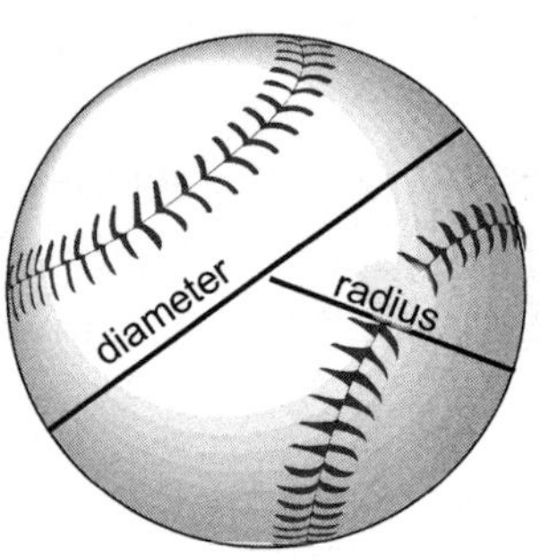

Fred drew on a baseball

Fred said, "Those are the same thing. Millimeters are abbreviated as mm."

Fred pointed to a metric poster that was taped to the side of the chem lab table.

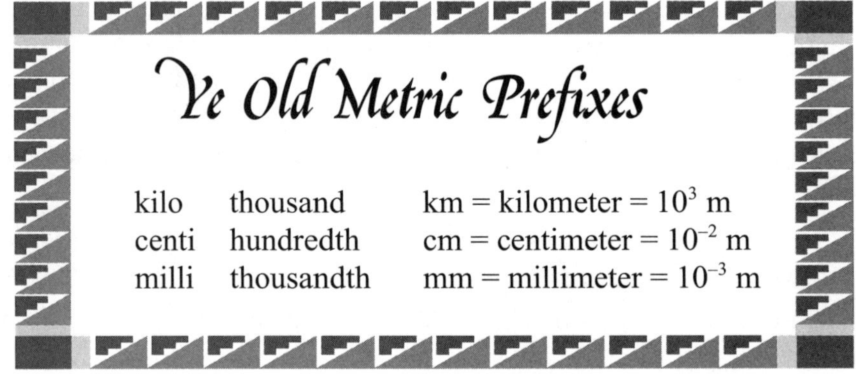
Ye Old Metric Prefixes

kilo	thousand	km = kilometer = 10^3 m
centi	hundredth	cm = centimeter = 10^{-2} m
milli	thousandth	mm = millimeter = 10^{-3} m

"But the handiest one for chem is still 1 Å = 10^{-10} m. It's perfect for measuring the diameters of atoms. Virtually all atoms are between one and five angstroms.

"Now I need to find out how many 2.88 Å (the diameter of gold atoms) are in a 0.5 mm pencil line.

"When you don't know whether to add, subtract, multiply, or divide, you use some simple numbers and you can see what operation to use. If I want to know how many 5-inch balls could line up to make a 20-inch width, I know the answer is 4. I divided.

"So I divide 2.88 Å into 0.5 mm." Fred wrote on the board . . .

$$\frac{0.5 \text{ mm}}{2.88\text{Å}} = \frac{0.5 \times 10^{-3} \text{ m}}{2.88 \times 10^{-10} \text{m}} = 0.1736 \times 10^{7} = 1{,}736{,}000$$

which rounds off to 2,000,000 gold atoms lined up side-by-side to make the width of a 0.5 mm pencil line.

"To give you some idea, how big two million is, let's consider two million inches. Let's convert that to feet using a **conversion factor**.

"Since 12 inches equals 1 foot, the conversion factor is either

$$\frac{12 \text{ inches}}{1 \text{ foot}} \quad \text{or it is} \quad \frac{1 \text{ foot}}{12 \text{ inches}}$$

"The top and bottom of a conversion factor must always be equal. Then when you multiply by a conversion factor, you are multiplying by a fraction that is equal to one. Multiplying by one doesn't change anything. Multiplying by a conversion factor will just change the units—in this case from two million *inches* into *feet*.

$$\frac{2{,}000{,}000 \text{ inches}}{1} \times \frac{1 \text{ foot}}{12 \text{ inches}} \doteq 166{,}667 \text{ feet}$$

$\doteq$ means "equal after rounding"

Joe raised his hand. "How do you know which conversion factor to use?"

Fred said, "I wanted to get rid of inches, so I put the inches on the bottom so that they would cancel."

$$\frac{2{,}000{,}000 \cancel{\text{ inches}}}{1} \times \frac{1 \text{ foot}}{12 \cancel{\text{ inches}}} \doteq 166{,}667 \text{ feet}$$

"This turned 2,000,000 inches into feet.

"But 166,667 feet is hard to visualize. I'm going to turn feet into miles. What is the conversion factor?"

One of the chem students went to the board and wrote . . .

$$\frac{5{,}280 \text{ feet}}{1 \text{ mile}} \quad \text{or it is} \quad \frac{1 \text{ mile}}{5{,}280 \text{ feet}}$$

Fred asked the student to finish it up by putting feet on the bottom so that the feet would cancel.

$$\frac{166{,}667 \text{ feet}}{1} \times \frac{1 \text{ mile}}{5{,}280 \text{ feet}} \doteq 31.6 \text{ miles}$$

"So if each gold atom were an inch wide, then two million of them would stretch for about 30 miles. Two million is a big number."

"In math, we don't have to do rounding. If we say seven, we mean exactly seven. 7 is the same as 7.0000000000000000000.

Bob shouted from the back of the room, "But when we do measurements in chemistry, we have to pay attention to significant digits and rounding."

Significant Digits Lesson

You start counting with the first non-zero number, which I'll underline. 703.09 0005987.90 0.00042703

You stop counting with the last non-zero number or the last gratuitous zero, whichever occurs last, which I'll double underline.

8201.003 2960000 9705.87000

A gratuitous zero is a zero that didn't have to be written. The zeros in 2960000 are *not* gratuitous. Without them, the number would be 296, which is an entirely different number. The three zeros at the end of 9705.87000 are gratuitous. Omit them, and the number doesn't change.

How Much to Round

Don't round in the middle of a computation. Keep all the digits. When you get to the final answer (and the computation was multiplications and divisions), then round your answer to the smallest number of significant digits in the given problem.

$$\frac{138.0}{5555} \times \frac{7.003}{9810} \times \frac{622222}{39275} = 0.000280956646711103289914249456118$$

$\doteq 0.000281$ **three significant digits since 9810 has three.**

I have a big calculator. That's how I could get 0.00028095664671110328991424945611 8. If you have one that can give only eight digits, that's okay. After you round, you will get exactly the same answer as I got.

Your Turn to Play

1. Some mechanical pencils use lead that has a diameter of 0.7 mm. How many carbon atoms lined up in a row would equal the width of a 0.7 mm line? The atomic radius of a carbon atom is 0.77 Å.

Round your final answer to the appropriate number of significant digits.

2. Last year the KITTENS University president made eighty kilodollars. How much is that?

3. If you have a ruler marked in metric units, you'll find that 5 centimeters equals approximately 2 inches. A centimeter is about the width of which of your fingernails?

4. S is an atom. H_2 is a ________________.
fill in one word

5. Using a conversion factor, calculate how long 9.3 inches is in centimeters. Note: 2 inches ≈ 5 cm

≈ means "approximately equal to."

6. When Joe went fishing last week, he caught what he thought was a giant worm. It was an eel.

It was 4' 9" long. (That's 4.75 feet.)

First change that to inches using a conversion factor. (1 foot = 12 inches).

Then change it to centimeters. (2 inches ≈ 5 cm)

This eel weighed 45 grams for each centimeter of its length. (1 cm = 45 g)

Joe was paid 4¢ for each gram.

How much did he receive?

.......COMPLETE SOLUTIONS.......

1. The diameter of a carbon atom is 1.54 Å. (The math: $2 \times 0.77 = 1.54$) We want to find out how many 1.54 Å are in 0.7 mm.

$$\frac{0.7 \text{ mm}}{1.54 \text{ Å}} = \frac{0.7 \times 10^{-3} \text{ m}}{1.54 \times 10^{-10} \text{ m}} = 0.45454545 \times 10^{7} = 4{,}545{,}454$$

$\doteq 5{,}000{,}000$ We round to one significant digit since 0.7 has only one significant digit.

2. \$80,000
3. For many people, their pinky fingernail is about a centimeter wide.
4. S is an atom. H_2 is a molecule.
5. The conversion factor will be either $\frac{2 \text{ inches}}{5 \text{ cm}}$ or $\frac{5 \text{ cm}}{2 \text{ inches}}$

$$\frac{9.3 \text{ inches}}{1} \times \frac{5 \text{ cm}}{2 \text{ inches}} = 23.25 \text{ cm} \doteq 23 \text{ cm}$$

6. Why do them one at a time when you can do them all in one (giant) step?

$$\frac{4.75 \text{ feet}}{1} \times \frac{12 \text{ inches}}{1 \text{ foot}} \times \frac{5 \text{ cm}}{2 \text{ inches}} \times \frac{45 \text{ g}}{1 \text{ cm}} \times \frac{4¢}{1 \text{ g}}$$

$= 25650¢ = \$256.50 \doteq \$300.$

Joe decided that instead of being called a *fisher*man, he wanted to be called an *eeler*man. That one eel was worth about half of Fred's monthly salary of \$600.

The only problem was that when Joe said he was an eelerman, no one knew what he was talking about.

Chapter Four
Weighing an Atom

Fred heard a slurping sound. He looked at you-know-who. Joe was working on the largest ice cream cone that Fred had ever seen. It must have been a foot tall.

Joe smiled (with ice cream on his face) and said, "Those jelly beans off the floor didn't taste so good." He pointed to the cone, "It's peanut butter and pineapple ice cream. Wanna lick?"

Fred shook his head.*

Joe raised his bandaged hand. The other hand held the cone. "You told us that virtually all atoms are between one and five inches wide. . . ."

"Angstroms, not inches," Fred interjected.

Joe continued, "Angstroms, whatever. But how much do those little guys weigh?"

Fred looked at the chem equipment on the lab table: flasks, Bunsen burners, tubing, a cylinder of helium gas, a scale that could weigh to the nearest thousandth of a gram, bottles of distilled water, acids, bases, beakers, and containers of many solid chemicals.

"You want me to find the weight an atom?" Fred asked.

"Sure. Why not?" Joe answered.

Bob shouted to Joe, "Hey, Joe. You are asking the impossible! Fred doesn't have the equipment to weigh an atom."

"I will do it."

The class was utterly silent. Even Joe stopped slurping.

These four words are, perhaps, the most famous four words ever uttered in any chemistry textbook.

---

* Even years later, Fred never figured out how Joe was able to do this trick—produce an ice cream cone part way through an hour-long class.

"I will find the weight of a single atom of helium using only the stuff on this lab table."

Joe didn't even ask his question, "Atoms—aren't they really light?"

Bob was shaking his head in disbelief. Bob knew that Fred was good at math, but *weighing an atom*?"

Fred began, "I will find that the weight of a single helium atom using this equipment."

Six of the students grabbed their phones and contacted their friends and the news agencies.

While Fred was clearing away some of the unneeded lab equipment, his auditorium classroom—all two thousand seats—were filled. Newspaper reporters, photographers, and cameramen had arrived.

Some of the sportscasters were set to give the play-by-play action as it unfolded.

Around the world, television programs were interrupted in order to cover this epochal event.

Everyone believed that Fred could do it. No one believed that it could be done. *Great truths in life are often enclosed in these kind of paradoxes.*

Fred found a glass container on the lab table. It was in the shape of a cube. He measured each dimension. Each side was 4 inches.

2 inches ≈ 5 cm (from #3 in the previous Your Turn to Play)

So each side was 10 cm.
The volume was 1,000 cm^3 (1,000 cubic centimeters).

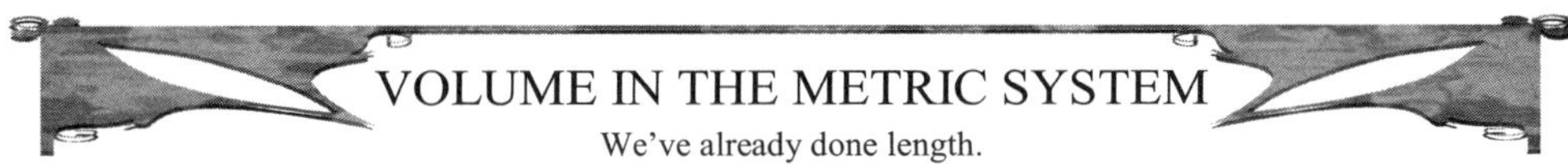

VOLUME IN THE METRIC SYSTEM

We've already done length.

The meter is one of the standard base units in the metric system.* Once you have defined a meter, you automatically get volume.

One centimeter is one hundredth of a meter.

A cubic centimeter (cm^3) is the volume of a cube that is one centimeter on each side.

1000 cm^3 is defined to be a **liter**.

That makes a cubic centimeter equal to a milliliter.

Sometimes a cubic centimeter is written a mL.

Sometimes a cubic centimeter is written as cc.

$1\ cm^3 = 1\ cc = 1 mL$

A liter is a little larger than a quart.
A meter is a little larger than a yard.

Fred picked up the 1 L glass container and placed it on the scale. Everyone cheered. They didn't know why they were cheering. The excitement was contagious.

It weighed 53.285 grams.

He attached a rubber hose to the helium container and turned it on. He blew out the air and replaced it with helium.

It now weighed 52.179 grams.

The sportscaster shouted into his microphone, "Fred has just proved that helium atoms weigh less than nothing." He had forgotten that balloons filled with helium float upward.

Bob Bunsen told the newspaper reporters who were near him, "All that shows is that helium atoms weigh less than the air that used to be in the glass container. $52.179 < 53.285$

---

* The other ones are: mass (gram), time (second), temperature (Kelvin), amount (mole), electric current (ampere), and luminous intensity (candela).

Fred wanted to weigh the helium in that 1 L container. If he knew the weight of the container, he could just subtract that from the weight of the container + the weight of the He. (← which he already knows is 53.285 g)

How to find the weight of just the container without the air in it?

"Hey, Bob," Fred said. "Do you have a vacuum pump?"

"Sorry, Fred."

"Bob, do you know any way to get the air out of this container?"

"Sorry, Fred."

"Bob, do you have any ideas?"

"Sorry, Fred. This is your show. I'm not the one who claimed he could weigh a helium atom."

"I never said that," Fred said. "I said that I could use this equipment to find the weight of a helium atom."

Everyone, except Fred, was confused.

He was merely worried that everyone in the world who was watching him on television would see him fail. Fred had failed many times in his life but never in front of billions of people.

What made it even worse was that he had said those famous words, "I will do it." (← find the weight of a helium atom)

The one fact that Fred was basing his whole experiment on was the law that is in every beginning chemistry book: *22.4 liters of any gas contains* 6.02×10^{23} *atoms (or molecules).*

Fred had hoped to find the weight of one liter of helium. Then he could multiply it by 22.4 and find the weight of 6.02×10^{23} atoms.

The divide by 6.02×10^{23} and get the weight of one atom.

But he was stuck at the first step. He couldn't get the weight of one liter of helium.

Unfortunately, we have to take a break for . . .

Your Turn to Play

1. Some gases are atoms, such as helium (He). Some are molecules, such as oxygen (O_2). In either case the number of particles—atoms or molecules—in 22.4 L of gas is 6.02×10^{23}. (That is the conversion factor.)

How many particles of hydrogen gas are in 2 L?

2. How many milliliters are in 86 cm^3?

3. Originally, the peanut butter and pineapple ice cream on Joe's cone was in the shape of a sphere (a ball). The diameter was 8 inches.

A) How many centimeters is that? (On page 32, page 30, and three times on page 29, we noted that 2 inches ≈ 5 cm.)*

B) What is the radius (in cm) of that sphere?

C) What is the volume of that sphere? (The formula for the volume of a sphere is $\frac{4}{3}\pi r^3$.) $\pi \approx 3.1415926535897932384626433832795$. Use as many digits of that as you like, but remember that you will be rounding your answer (because of significant digits).

4. If the diameter of a mechanical pencil lead is 0.5 mm, then when you draw a long line, the width of the line should be 0.5 mm. How wide is that in inches?

Data for conversion factors: 1 mm = 10^{-3} m (exactly)
1 cm = 10^{-2} m (exactly)
2 inches ≈ 5 cm

---

* The end of this sentence ". . . we noted that 2 inches ≈ 5 cm.)" has a period. That's because it is the end of a sentence.

The English system sometimes uses periods: ft or ft. and lbs or lbs.

The metric system does not use periods: cm or m or mL in writing the abbreviations.

.......COMPLETE SOLUTIONS.......

1. The conversion factor will either be

$$\frac{22.4\text{ L}}{6.02 \times 10^{23}\text{ particles}} \quad \text{or it will be} \quad \frac{6.02 \times 10^{23}\text{ particles}}{22.4\text{ L}}$$

We start with the given 2 L and want to convert that into particles.

$$\frac{2\text{ L}}{1} \times \frac{6.02 \times 10^{23}\text{ particles}}{22.4\text{ L}} = 0.5375 \times 10^{23}\text{ particles}$$

$\doteq 0.5 \times 10^{23}$ particles (since 2 L has only one significant digit)

which is equal to 5×10^{22} in **scientific notation**. (Numbers in scientific notation are in the form $n \times 10^m$ where $1 \leq n < 10$.)

2. This is silly. 86 milliliters $1\text{ cm}^3 = 1\text{ cc} = 1\text{mL}$

3A) $\frac{8\text{ inches}}{1} \times \frac{5\text{ cm}}{2\text{ inches}} = 20$ cm is the diameter

3B) 10 cm is the radius. (A radius is half of the diameter.)

3C) volume $= \frac{4}{3}\pi r^3 = \frac{4}{3}(3.14159265)10^3 = 4188.79$

$\doteq 4{,}000\text{ cm}^3$ (since 10 has only one significant digit) or 4,000 cc or 4,000 mL = 4 L

4. We start with the given 0.5 mm.

$$\frac{0.5\text{ mm}}{1} \times \frac{10^{-3}\text{ m}}{1\text{ mm}} \times \frac{1\text{ cm}}{10^{-2}\text{ m}} \times \frac{2\text{ inches}}{5\text{ cm}} = 0.02\text{ inches}$$

Joe once went **cancel crazy**. He was working with the conversion factor $\frac{100\text{ cm}}{1\text{ m}}$ and he cancelled the m's and got $\frac{100\text{ c}}{1}$ and then he cancelled the 1s and got $\frac{00\text{ c}}{}$

Chapter Five
On to Victory

Fred could be creative—especially when seven billion people and one monkey are watching. He needed to get the weight of an empty 1 L glass container and ultimately find the weight of 1 L of helium.

Fred's First Thought: There is no air in outer space. All I have to do is send the glass container and the scale into outer space where there is an (almost) perfect vacuum.

Objection #1: Bob's chem lab table didn't have any rocket ships.

Objection #2: Even if it did, Fred would have to also go on the ship so that he could read the scale. He didn't want to risk his life for a chemistry experiment.

Answer to Objection #2: Fred could send a robot to read the scale and report back what he read.

Objection #3: In outer space things just float. There is virtually no gravity. The scale would read 0 grams.

MASS vs. WEIGHT

When you step on the scale and it reads 142 pounds, that is a weight (a force).

Mass is not a force or a weight. Mass is the measure of how much stuff–how much matter–is in something. If Joe's bag of jelly beans is 4 kilograms on the surface of the earth, it will be 4 kg in outer space.

When you plop that bag of jelly beans on a chemistry scale, it might read 4000 g. This is a lie. Well . . . sort of. All a scale can measure is a weight (a force)–such as pounds or ounces. If you took that scale and the bag of jelly beans up several hundred miles above the surface of the earth (where gravity is less), the scale might read 3285 g. That would be false. The bag would still have a mass of 4000 g. You didn't lose any of the molecules.

When they build those scales that give readings in grams, they make the hidden assumption that you are using them on the surface of the earth. But they never tell you that.

Those scales (when you are not looking) actually measure the force (pounds) and use the conversion factor $\frac{1000 \text{ g}}{2.205 \text{ lbs.}}$ since 1 kg ≐ 2.205 lbs. on the surface of the earth.

Those sneaky scales!

Fred's First Thought wouldn't work.

Fred's Second Thought: I need to get rid of the air that's in the 1 L glass container and then weigh it. What if I fill that container with something that will displace the air and for which I know the weight?

Water! Everyone who's had chem knows that 1 mL of water weighs 1 g.*

Fred flew into action. He filled the glass container with water and weighed it. 1,052.000 g

1 L of water equals 1000 mL of water, which has a weight of 1000 grams.

Subtracting (1052.000 – 1000.000) and we find that the truly empty glass container weighs 52.000 grams.

The rest would be a walk in the park (an idiom). It would be as easy as pie (another idiom). It would just be simple math—subtraction and multiplying conversion factors.

---

* Let's not be so picky and say that the water in one cubic centimeter has a *mass* of 1 gram.

Some boxer might say, "I *weigh* 140 kilograms." You don't correct him and say, "You have a *mass* of 140 kilograms." You would be more correct, but you don't want to anger a boxer who weighs more than 300 pounds. $\frac{140 \text{ kg}}{1} \times \frac{2.205 \text{ lbs.}}{1 \text{ kg}} = 308.7 \text{ lbs.} \doteq 310 \text{ lbs.}$

Fred wrote on the board . . .

Weight of truly empty 1 L glass container (no air in it)	52.000 g
Weight of container + 1 L of He*	52.178 g
Weight of 1 L of He (by subtraction)	0.178 g

Using conversion factors to change 1 atom of He into its mass . . .

$$\frac{1 \text{ atom of He}}{1} \times \frac{22.4 \text{ L}}{6.02 \times 10^{23} \text{ atoms}} \times \frac{0.178 \text{ g}}{1 \text{ L}}$$

equals . . . the audience waited as Fred (and many of his students) punched it out on their calculators . . .

$0.66232558139534883720930232558{14} \times 10^{-23}$ grams

$\doteq 0.000\,000\,000\,000\,000\,000\,000\,006\,62$ grams

"I did it."

These three words are, perhaps, the most famous three words ever uttered in any chemistry textbook.

Fred had used some basic chem lab equipment and two elementary facts: 1 mL of water weighs 1 g and *22.4 liters of any gas contains 6.02 × 10^{23} particles (atoms or molecules).*

Fred did a quick computation in his head. A 22.4 L container would be a little over 11 inches on each side.

Joe had eaten so much that he had fallen asleep. The students who had been sitting next to him had moved away. They

---

* Fred made this weighing back on page 31.

didn't want to get any sticky peanut butter and pineapple ice cream on them.

One of the newspaper reporters had a question. "I don't get it. What if you vacuumed out some of those 6.02×10^{23} atoms of helium out of the 22.4 L container? The volume of the container wouldn't change, but you'd have less than 6.02×10^{23} atoms of helium left inside."

That made a lot of sense.

One of the students said, "Or what if you pumped in some more helium atoms under higher pressure? Then the 22.4 L container would have more than 6.02×10^{23} atoms of helium in it."

That also made a lot of sense.

Fred said, "But that is what I read in all the chemistry textbooks: *22.4 liters of any gas contains 6.02×10^{23} particles (atoms or molecules).*"

Bob Bunsen came to Fred's rescue, "That's true, but only under **standard pressure**." Fred had missed that little bit of information.

Standard pressure is 14.7 pounds per square inch (abbreviated psi).

Standard pressure is one atmosphere (abbreviated atm—not to be confused with a bank's ATM).

Standard pressure is 760 mm of mercury. Mercury (symbol Hg) is a metal that is liquid at room temperature. It has a shiny silver color. A column of Hg that is 76 cm (or 760 mm) tall would exert the same pressure as a column of air does that goes from ground level up to outer space. You don't feel the weight of all that air because your body is pushing back with the same force. When you drive up into the high mountains, there is less pressure and your ears pop. They are adjusting to the decrease in pressure.

When they take your blood pressure and it's 120/80, that means that the pressure is varying between 120 mm of Hg and 80 mm of Hg. It is varying because your heart is beating (which is nice). Heart pushes: 120; heart relaxes: 80.

Fred quickly put a pressure gauge on his 1 L sample of helium. It read 760 mm of Hg. He was so relieved. His 0.000 000 000 000 000 000 000 006 62 grams for the weight of a helium atom was correct.

Your Turn to Play

1. If you double the pressure, you will double the number of particles of gas that you can pack into any given volume.

You have 6.02×10^{23} molecules of O_2 in a 22.4 L container at standard pressure.

If you pump in more oxygen, so that the pressure increases from 760 mm of Hg to 900 mm of Hg, how many molecules of O_2 will be in the container? (Use a conversion factor.)

2. You want some really low pressure? The number of molecules in a container is proportional to the pressure. Suppose we pump out all of the molecules of O_2 in a 22.4 L container and leave only 8 molecules. What would be the pressure (measured in millimeters of mercury)?

3. Let's test your creativity. You are in your kitchen and the recipe calls for one cup of syrup. Your sister took all the measuring cups and measuring spoons when she moved to Denver last week. You have nothing that measures volume. You do have scales and thermometers and chem books and spoons and all the other usual stuff that is in a kitchen.

You don't know how much a cup of syrup weighs.

Your chem book says that a quart (which is 4 cups) equals 0.9463 liters.

How might you measure that cup of syrup (without going to the store and buying some new measuring cups)?

Traditional chem textbooks contain 60 problems at the end of each chapter. And they are mostly just drill. They ask the same question over and over again, just changing the numbers. **Perhaps, I exaggerate a little.**

No other chem book talks about having only eight molecules of gas or about your sister moving to Denver.

These three puzzles in this *Your Turn to Play* may take as much time as 60 drill problems, but they can be much more fun.

.......COMPLETE SOLUTIONS.......

1. We start with 6.02×10^{23} molecules of O_2 at 760 mm (standard pressure). To create the conversion factor we need to find two quantities that are equivalent: the number of molecules at 760 mm and the number of molecules at 900 mm.

$$\frac{6.02 \times 10^{23} \text{ molecules of } O_2 \text{ at 760 mm}}{1} \times \frac{\text{no. of molecules at 900 mm}}{\text{no. of molecules at 760 mm}}$$

$= 7.13 \times 10^{23}$ molecules at 900 mm

2. We start with a pressure of 760 mm with 6.02×10^{23} molecules of O_2 in the container.

$$\frac{760 \text{ mm}}{1} \times \frac{8 \text{ molecules}}{6.02 \times 10^{23} \text{ molecules}} = 1009.9 \times 10^{-23} \text{ mm}$$

$\doteq 1.01 \times 10^{-20}$ mm of Hg

3. Here is one way. Your way might be different.
   Find out how much a cup of water weighs.*
   Fill a glass with 237 grams of water (using your scale).**
   Mark the glass at the top of the water level.
   You now have a one-cup measuring cup.

---

* $\frac{1 \text{ cup}}{1} \times \frac{1 \text{ quart}}{4 \text{ cups}} \times \frac{0.9463 \text{ L}}{1 \text{ quart}} \times \frac{1000 \text{ g of water}}{1 \text{ L}} \doteq 237 \text{ g of water}$

(I love playing with conversion factors!)

** You weigh the empty glass and add water until it's 237 grams heavier than the starting weight.

Chapter Six
Hotter and Colder

Fred thought about his experiment to find the weight of a helium atom (0.000 000 000 000 000 000 000 006 62 grams). Now that he knew about standard pressure, he figured out a way to do the whole thing without learning the weight of the empty 1 L glass container.

The newspaper reporters had already closed their notebooks. The cameramen had already started packing up their equipment.

Fred announced, "You haven't seen anything yet. No chemistry textbook on earth has ever found the weight of a helium atom without using water and without learning the weight of the container holding the helium."*

The notebooks were opened. The cameras were unpacked. Bob Bunsen had no idea how Fred was going to do this "miracle".

Fred took the container that had been filled with helium at standard pressure (one atm) and pumped in more helium until the pressure was two atmospheres.

Fred was wearing safety glasses. He didn't want to be blinded if the glass shattered. Everyone in the front two rows moved to the back of the auditorium.

With one liter of He the container weighed	52.178 g
With two liters of He it weighed	52.351 g

Subtract (The math: 52.351 – 52.178 = 0.173) and he announced that the weight of one liter of He was 0.173 grams.

Rats! He was expecting a liter of helium to weigh 0.178 grams. That's what he had found using the water approach.

He rechecked the pressure. Oops. It was only 1.9 atm. He pumped in some more helium until the pressure was 2 atm.

---

* *Life of Fred: Chemistry* was written after all these events had taken place.

He reweighed. It was 52.355 grams. One liter of helium was 0.177 grams. (The math: 52.335 – 52.178 = 0.177). It still wasn't up to the 0.178 grams when he used the water approach.

He checked the pressure again. It was only 1.96 atm.

Obviously, Fred thought there must be a leak in the equipment. When I teach math, my numbers don't leak on the blackboard. And I don't have to wear safety glasses either.

"Bob," Fred said. "Can you help me find the leak in your equipment? The pressure keeps going down when I fill up the container."

Bob smiled. "There's no leak. But of course, it's going down. That's to be expected."

"Expected! Do you expect that there's going to be a pressure drop after you pump the gas into the container?"

Some background: Fred hadn't anticipated teaching chemistry today. At nine o'clock was his beginning algebra class and at ten was his advanced algebra class. Things had changed when Joe, a bloody mess, had staggered into his classroom. Somehow, now, he was standing at a lab table teaching chem and being watched by seven billion people.

Fred knew some of the elementary stuff: a milliliter of water weighs one gram. At standard pressure, 22.4 liters of any gas contained 6.02×10^{23} particles.

He rechecked the pressure. It had been 1.96 atm. It was now down to 1.95 atm. And Bob said that nothing was leaking.

The monkey that was watching Fred's distress on television began to laugh again. Luckily, Fred couldn't hear him.

Ha! Ha! Ha!

In desperation he pumped in helium until the pressure was 2.5 atm. He watched as the pressure started to drop 2.43 atm, 2.42 atm, 2.39 atm. He put his hand on the top of the container. It was hot.

As it was cooling off, the pressure was dropping.

He opened one of the chem manuals that was on the lab table. This time he read it slowly: *At standard temperature and pressure, 22.4 liters of any gas contains 6.02×10^{23} particles.* (← This time it is true.)

Standard temperature and pressure (which everyone abbreviates as STP) is 0° C (= 32° F) and 1 atmosphere (= 760 mm of mercury). The "C" stands for the Celsius temperature scale.

Again, like meters and grams and liters, the Celsius scale is used almost everywhere in the world . . . except in the United States and two other countries.

Here are some handy reference points so you can make the transition from Fahrenheit to Celsius:

Water freezes at 32° F = 0° C
A cool room 68° F = 20° C
A warm room 77° F = 25° C
A body temperature of 98.6 = 37° C
Water boils 212° F = 100° C

Now Fred could find the weight of a helium atom using the whole truth: *At STP, 22.4 liters of any gas contains* 6.02×10^{23} *particles.*

Armed with a thermometer and some ice (from Joe's ice chest where he kept his bottles of Sluice), Fred went to work.

He filled the 1 L container with helium cooling it down with ice and adding more helium until he had STP (0° C and 760 mm of Hg).

Weight of 1 L container + He at STP 52.179 g

He then added more helium and cooled it until he had a pressure of 2 atm and a temperature of 0° C.

Weight of 1 L container + He at 0° C and 2 atm 52.358 g

Weight of 1 L of He (by subtracting) 0.179 g

Using conversion factors to change 1 atom of He into its mass . . .

$$\frac{1 \text{ atom of He}}{1} \times \frac{22.4 \text{ L}}{6.02 \times 10^{23} \text{ atoms}} \times \frac{0.179 \text{ g}}{1 \text{ L}}$$

equals . . . the audience waited as Fred (and many of his students) punched it out on their calculators . . .

$0.66604651162790 \times 10^{-23}$ grams

$\doteq$ 0.000 000 000 000 000 000 000 006 66 grams

which was slightly different than the 0.00000000000000000000000662 grams he had computed when he was unaware that the measurements needed to be made at standard temperature and pressure.

Fred had found the weight of a single helium atom.

LET'S SIMPLIFY THINGS A BIT

Earlier, we got tired of writing that the radius of an atom of gold (Au) was 0.000000000144 meters.

We defined an angstrom (Å) as 10^{-10} m. That's one ten-billionth of a meter.

Then the radius of an atom of gold could be written as 1.44 Å. That's much easier on the eyes.

Chemists also got tired of writing 6.02×10^{23} or 602,000,000,000,000,000,000,000 when they wanted to describe the number of particles (atoms or molecules) in 22.4 liters of a gas at STP.

This number is known as **Avogadro's number**. But *Avogadro's number* has six syllables. That's a mouthful. Instead, chemists call 6.02×10^{23} a **mole**.

Egg farmers call 12 eggs a dozen.

Lincoln called 20 a score. ("Four score and seven . . .")

Paper manufacturers call 500 sheets a ream.

$17,000,000,000,000 = our national debt

602,000,000,000,000,000,000,000 = one mole

Joe dreamed of writing to Santa and asking for a mole of jelly beans. He thought that a mole of jelly beans would fill up his living room.

$$\frac{\text{1 mole of jelly beans}}{1} \times \frac{6.02 \times 10^{23}}{\text{1 mole}} \times \frac{\text{1 cubic inch}}{\text{50 beans}}$$

$= 0.1204 \times 10^{23}$ cubic inches

There are 12^3 cubic inches in a cubic foot. If you took a cubic foot and divided it into cubic inches, there would be $12 \times 12 \times 12$ of them.

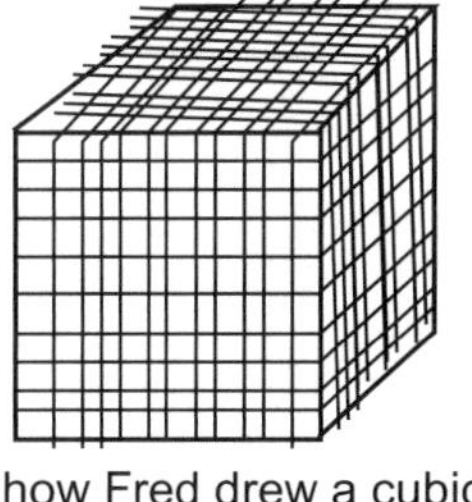
how Fred drew a cubic foot on the board

There are $5{,}280^3$ cubic feet in a cubic mile.

Continuing the computation of one mole of jelly beans . . .

$$\frac{0.1204 \times 10^{23} \text{ cubic inches}}{1} \times \frac{1 \text{ cubic foot}}{12^3 \text{ cubic inches}} \times \frac{1 \text{ cubic mile}}{5{,}280^3 \text{ cubic feet}}$$

$\doteq 4.73 \times 10^7$ cubic miles $= 47{,}300{,}000$ cubic miles, which is a cube whose edges are each longer than 361 miles. A mole of jelly beans would not fit in Santa's sleigh.

Moles are great for counting atoms, but not for jelly beans.

Your Turn to Play

1. If Fred had used some really good scales he would have found that the weight of an atom of helium was 0.66495×10^{-23} grams.

   What is the weight of a mole of helium atoms?

2. A molecule of oxygen (O_2) weighs 5.31561×10^{-23} grams. How much would a mole of O_2 weigh?

3. How much would a mole of oxygen atoms (O) weigh?

4. How much would a mole of ozone (O_3) weigh?

5. Chlorine (Cl_2) is a pale yellow-green gas. It is one of the most heavily manufactured gases in the world. (One way to get chlorine is to take ordinary table salt—NaCl, sodium cloride—and break it into sodium and chlorine. Later, we will write the equation $2\,NaCl \rightarrow 2\,Na + Cl_2$ where Na is a solid metal and Cl_2 is a gas, but this is too early in the book to mention that equation yet.)

   If this were a history book, we could mention the use of chlorine gas attacks in World War I. When the gas hits the water in lungs, it produces hydrochloric acid ($2\,Cl_2 + 2\,H_2O \rightarrow 4\,HCl + O_2$). This is not good.

   What volume would 6 moles of Cl_2 occupy at STP?

.......COMPLETE SOLUTIONS.......

1. We want to convert one mole of helium into a weight. We start with one mole of helium. That's logical.

$$\frac{1\ \text{mole of He}}{1} \times \frac{6.02 \times 10^{23}\ \text{atoms}}{1\ \text{mole}} \times \frac{0.66495 \times 10^{-23}\ \text{grams}}{1\ \text{atom of He}}$$

$\doteq$ 4.003 grams

Moles are great. We can psychologically deal with 4.003 grams. A teaspoon of sugar weighs about 4 grams. Two pennies weigh about 5 grams.

2. $$\frac{1\ \text{mole of } O_2}{1} \times \frac{6.02 \times 10^{23}\ \text{atoms}}{1\ \text{mole}} \times \frac{5.31561 \times 10^{-23}\ \text{grams}}{1\ \text{atom of } O_2}$$

$\doteq$ 32.000 grams

3. Half as much as a mole of O_2: 16.000 grams
4. Three times as much as a mole of O: 48.000 grams

5. We start with 6 moles of Cl_2 and want to convert that into a volume.

$$\frac{6\ \text{moles of } Cl_2 \text{ at STP}}{1} \times \frac{22.4\ \text{L}}{1\ \text{mole at STP}} = 134.4\ \text{L}$$

$\doteq$ 100 L when we round off to one significant digit.

If the original problem asked us to convert 6.00 moles of chlorine, our answer would have been 134 liters.

Translation of $2\,Cl_2 + 2\,H_2O \rightarrow 4\,HCl + O_2$

Two molecules of chlorine gas combined with two molecules of water

and produce ($\rightarrow$)

four molecules of hydrochloric acid and one molecule of oxygen.

Chapter Seven
Asking Nature

Fred had already done two ground-breaking procedures in chemistry. First, he had determined the weight of an atom of helium. Second, he had done that without learning the weight of the one liter container.

He knew that if you doubled the pressure, you could stick twice as many particles (atoms or molecules) into any given volume.*

He knew that for a given volume, if you heat up the gas it will increase the pressure.

Fred wondered **If I double the temperature, will it double the pressure?**

small essay

The Scientific Method

In the old days—and nowadays—we start by getting a hunch. For example: *If I double the temperature of a gas, I bet that will double the pressure.*

Hunches are the way we start on the road to truth. We make a guess.

There are three ways to proceed from hunch to truth:
✪ Ask yourself. ✪ Ask an authority. ✪ Ask nature.

Ask yourself. Think about what is most probably true. Make up a story that fits your hunch. Tell a tale, such as there are four elements out of which everything is made: earth, air, fire, and water. Say that it seems logical that everything in the heavens is perfect.

Ask an authority. In Galileo's time, the authorities declared that everything in the heavens is perfect and that the sun revolved around the earth. Anyone with half-way decent eyeballs could see that the moon was

---

* In math, we say that the number of particles is proportional to the pressure (for a given volume).

blotchy rather than perfectly smooth. Did an authority figure (Mom) ever tell you about this fat guy who comes down chimneys and leaves presents?

Ask nature. That means go and do experiments, lots of them. Rather than trying to tell nature what must be true, let it tell you what is true. This is called the scientific method.

In 1919 during a solar eclipse, scientists were told by nature that gravity bends space. Since then, nature has told us that many of the things that we thought were true are just plain wrong.

end of small essay

If Fred asked Joe when he was awake, "If I double the temperature, will that double the pressure?" Joe would probably say, "Sure. That makes sense."

If Fred asked Bob Bunsen, he would probably respond, "Sure. That's true." Fred might have missed the rest of Bob's answer: "Sure it's true but only if you use the right..." It's not very good if your chemistry teacher mumbles.

Fred was going to use the scientific method. He was going to ask nature by doing an experiment.

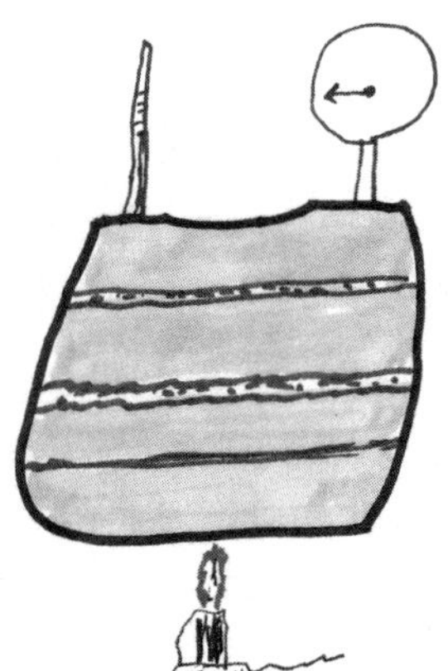

He located a metal container with a pressure gauge and thermometer on it. He didn't want to use the one liter glass container, because (1) it might melt if he applied a Bunsen burner flame to it, and (2) it might shatter if the pressure got too high.

First measurement: 23° C (room temperature) and the pressure was 760 mm of mercury. It didn't matter what gas was inside.

Apply a bit of heat using a Bunsen burner.

Second measurement: 46° C, and the pressure was 819 mm of Hg.

The pressure had gone up, but it didn't double when the temperature doubled. It didn't do what it logically should have done.

More heat. Third measurement: 57° C, and the pressure, 847 mm.

Being a mathematician, Fred saw three points, (23°, 760), (46°, 819), and (57°, 847), and he plotted them.

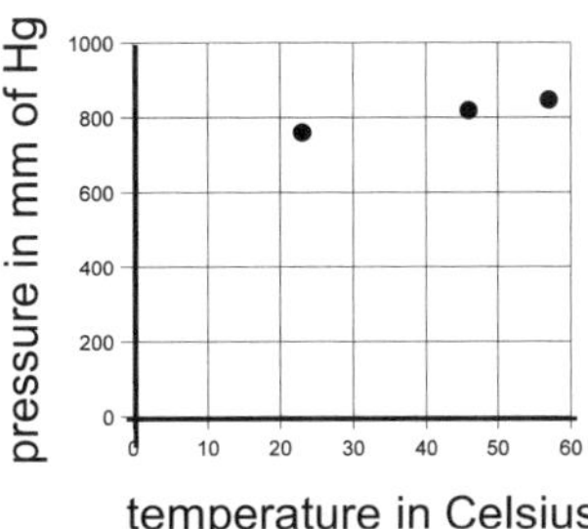

Fred gasped and thought **The points are in a straight line!**

He redrew the graph and drew a line through the three points.

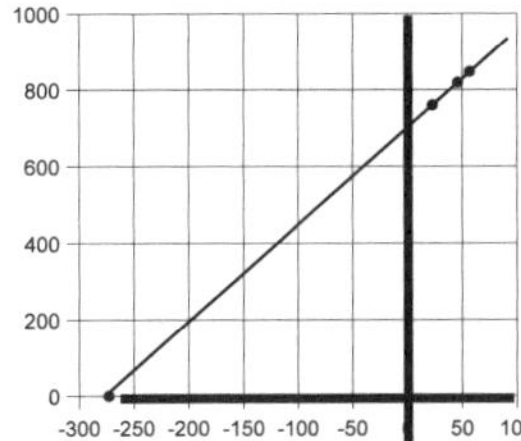

With his eagle eyes, he saw that the line hit the x-axis (the temperature scale) at –273° C.

Bob Bunsen stood up and applauded. "You did it! You are the first chem teacher to find the value of **absolute zero** just using elementary chem lab equipment. I didn't think that could be done. Over the years I've just told my students that –273.15° C is absolute zero. They had to take it on faith. You asked nature and she answered you."

Fred wasn't quite sure what Bob was talking about. "Could you please explain for the benefit of everyone in the room what this absolute zero thing is."

Around the world the television image was→ The image wasn't very complimentary.

Fred reaches absolute zero

←Zero

The reporters were reporting about science that they didn't understand.

Bob explained, "Molecules in any substance are always wiggling. If you increase the temperature, they wiggle more. Heat up an ice cube and the molecules of H_2O vibrate so hard that the ice turns to water. Heat up water and at 100° C the water turns to steam.

"On the other hand, cool air down far enough and the two major gases in air—oxygen and nitrogen—will turn into a liquid.

"At absolute zero, –273.15° C, all the wiggling stops.

"As I said before, if you double the temperature, you double the pressure, but that is true only if you are measuring from absolute zero."

A NEW TEMPERATURE SCALE

When we work with gases and measure pressure and temperature, we need a temperature scale that measures from absolute zero (–273.15° C) rather than from the melting point of ice (0° C).

The new scale is called the **absolute temperature scale**. (For once, scientists thought of the perfect name for something.)

It is also called the Kelvin scale because a guy named Lord Kelvin suggested the absolute temperature scale back in 1848.

The math is easy. Take a temperature in Celsius and add 273.15 and you know the temperature in kelvins.

Since 0° C is 273.15 degrees warmer than absolute zero, in the Kelvin scale 0° C is 273.15 K.

A cool room (20° C) is 293.15 K. (The math: 20 + 273.15 = 293.15)

Wait! I, your reader, found an error. Mr. Author, you wrote 293.15 K and forgot the degree mark (°).

Scientists are not always logical when they name things. They say 68° F "sixty-eight degrees Fahrenheit" is equal to 20° C "twenty degrees Celsius" is equal to 293.15 K "two hundred ninety-three point one five kelvins."

You forgot to capitalize "kelvins"!

You write 293.15 K or you can write 293.15 kelvins. I know it's crazy. You just have to live with it. I still can't figure out why they spell yacht as y-a-c-h-t instead of yot.

The nice thing is that once you convert from Celsius to Kelvin, pressure and temperature are proportional. Translation: You can use conversion factors.

For example, in Fred's experiment when the temperature was 23° C, the pressure was 760 mm of Hg. We can now *predict* what the pressure will be at 46° C.

First convert the 23° C and 46° C to 296.15 K and 319.15 K.

We are starting with 760 mm.

$$\frac{760 \text{ mm}}{1} \times \frac{319.15 \text{ K}}{296.15 \text{ K}} \doteq 819 \text{ mm}$$ (which is what Fred found in his experiment.)

One last question. How do I know whether to write the conversion factor as $\frac{319.15\ K}{296.15\ K}$ or as $\frac{296.15\ K}{319.15\ K}$?

Rather than give you lots of rules or formulas, the easiest way is use what you already know. In the example I just gave, the temperature was going up, so I *expect* the pressure to go up. So I use the conversion factor with the bigger number on top.

Your Turn to Play

1. There is a mole of chlorine in a 22.4 L container at standard temperature and pressure. (0° C and 760 mm of Hg)

If the container is heated to 30° C, how many moles of chlorine will be in the container?

2. If there are 0.5 moles of helium in a container at STP and the container is heated to 30° C, what will be the pressure?

3. How many moles of hydrogen are in a 1 L container at STP?

4. Nitrogen gas (N_2) is a gas that you should be well acquainted with. Air is 78% nitrogen. When we take a breath, our lungs generally ignore the nitrogen and carefully look for the oxygen molecules (O_2) that comprise 21% of the volume of air. My chemistry teacher, years ago, explained that nitrogen is like the 78% of daytime television in that it can safely be disregarded.

Nitrogen, like oxygen, is colorless and odorless. If you have breathed lately, you knew that.

Suppose a container of N_2 is at STP. What will be the pressure if the temperature is changed to –15° C?

.......COMPLETE SOLUTIONS.......

1. Hee hee hee. Did you think that heating up the container would make you lose (or gain) molecules of the gas? None of them escaped. After heating, you still have a mole of them.

2. The fact that there are 0.5 moles of helium is irrelevant. In real life, the problems you encounter don't just contain only the information that you need. You might also know the price of helium, but that is irrelevant. So is the fact that the container was manufactured in Texas.

Many algebra books (and chem textbooks) have tons of exercises in which *only the needed information* is supplied. Readers then just pick out the numbers and their only question is, "Do I add, subtract, multiply, or divide?" Those readers are not being prepared for life in the real world.

The helium begins at 0° C and 760 mm pressure. We want the pressure at 30° C. We expect the pressure to increase. The conversion factor will have the big number on the top.

$$\frac{760\text{ mm}}{1} \times \frac{303.15\text{ K}}{273.15\text{ K}} \doteq 843\text{ mm of Hg.}$$

3. We start with a volume of one liter. We want to convert that to moles.

$$\frac{1\text{ L}}{1} \times \frac{1\text{ mole}}{22.4\text{ L}} = 0.0446428\text{ moles} \doteq 0.04\text{ moles}$$

4. We start with a pressure of 760 mm (at 0° C) and want to know the pressure at –15° C. Since the temperature is falling, we expect the pressure to fall. The conversion factor has the small number on top.

$$\frac{760\text{ mm}}{1} \times \frac{258.15\text{ K}}{273.15\text{ K}} \doteq 718\text{ mm}$$

(or 720 mm if you, correctly assume that 760 mm has two significant digits.)

Did you notice how delightfully little arithmetic was involved in doing these four problems? (I'm assuming you used a calculator.) It is much harder to compute $3\frac{1}{8} - 1\frac{3}{5}$

Memorizing "stuff" is not an education. A pencil and a piece of paper can memorize $\pi = 3.14159265358979323846264338327 95$ perfectly and will remember it 20 years from now. We are seeking *understanding.*

Chapter Eight
Lots of Room

Fred wrote on the board: *Any gas at standard temperature and pressure has Avogadro's number of particles (atoms or molecules) in 22.4 liters.* He debated whether he should write *Avogadro's number* or just *mole.* He crossed out *Avogadro's number* and wrote 6.02×10^{23}.

Joe woke up when Fred said, "Avogadro." He thought that might be an ice cream flavor. Avogadro's number was named after Amadeo Avogadro. His mother (Mrs. Avogadro) would have been really mad if she heard that some American student thought that their family name was an ice cream flavor. Amadeo was born in the same year (1776) that the Notice of Secession from England (also known as the Declaration of Independence) was signed.

It was a natural mistake on Joe's part. He thought Fred had said, "Avocado," which is one of Joe's favorite ice cream flavors.

Joe raised his bandaged hand. "I have a question."

Fred looked at Joe and just waited. Sometimes Joe's questions were very easy to answer and sometimes they were mind-bendingly difficult.

"I was reading what you wrote on the board."

Of course, this wasn't a question. Fred continued to wait.

"It doesn't make sense."

This also wasn't a question.

"You wrote *any gas.*"

That really didn't sound like the question, but Fred answered, "Yes, I did."

"You would think that a mole of really big atoms would take up much more room than a mole of skinny atoms. Six skinny guys can sit on a park bench, but six fat ones won't fit."

"Yes," Fred answered. "You might think that."

Then came Joe's question, "Would you show me why big atoms don't take up any more space than skinny atoms?"

Bob Bunsen was so glad he wasn't up in front of the classroom trying to answer that hard question. In all the years that Bob taught, he just asked his students to take on faith the fact that the *every gas at standard temperature and pressure has Avogadro's number of particles (atoms or molecules) in 22.4 liters.* He was the teacher and his students had to believe him.

Fred was about to answer the question that virtually no chemistry textbook ever answers: Why it doesn't matter whether the gas is made up of fat atoms or skinny atoms.

"If you think of atoms of a gas as people trying to crowd into an elevator," Fred began, "then, certainly, you could fit more skinny people in than fat people.

"But that is not exactly what it looks like when you stuff a mole of gas particles (atoms or molecules) into 22.4 liters of space.

"Virtually all atoms have diameters between one and five angstroms. (1 Å = 10^{-10} m.)

Joe raised his hand and said, "See. I was right. The fat atoms would take up five times as much space as the skinny ones."

Fred smiled. "That would be true if they were all tightly packed together. Let's do the math."

That was the signal for Joe to take another nap.

On the board, Fred wrote . . .

Take a medium-sized atom. Say gold.
Diameter of a gold atom = 2.88 Å.
Radius = 1.44 Å. (The math: $2.88 \div 2 = 1.44$)
From geometry, volume of a sphere (a ball) = $(4/3)\pi r^3$
Volume of a gold atom = 1.25×10^{-29} L
Volume of a mole of gold atoms = 7.53×10^{-6} L
These gold atoms are running around in 22.4 liters.

$$\frac{7.53 \times 10^{-6}\ \text{L}}{22.4\ \text{L}} \doteq 0.0000336\%$$

One of the reporters shouted to Fred, "Hey! Isn't 0.0000336% hugely small?"

The class grew very quiet. The English language had just been destroyed.

Fred said, "Let's just say that if a mole of gold atoms as a gas takes up only 0.0000336% of the volume of the container, then they aren't crowded together like people on a park bench. They have plenty of elbow room."

The KITTENS campus newspaper ran a special edition.

THE KITTEN Caboodle

The Official Campus Newspaper of KITTENS University 10:32 a.m. Edition 10¢

exclusive

Fred Proves that Gold Leads Lonely Life

KANSAS: If you were a gold atom in a gas of gold atoms, you wouldn't be constantly running into your friends.

Imagine two blind people in a large room and each one throws a tennis ball in any direction. What are the chances those balls would collide?

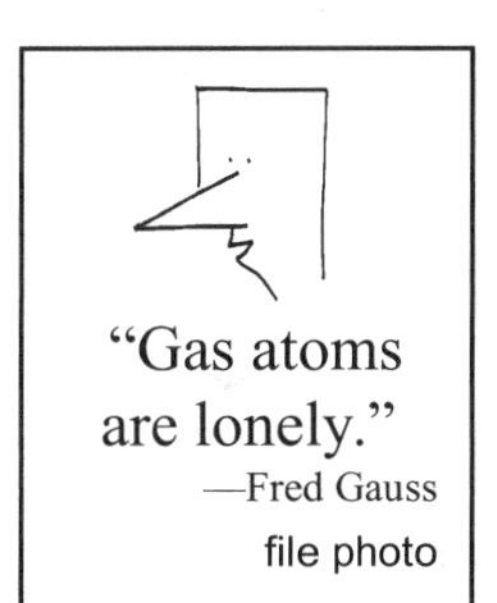

file photo

One of the students handed Fred a copy of the newspaper. Fred looked at it. At least they got the photo right Fred thought to himself. Once they ran a picture of me that looked like this:

But their science reporting isn't very good. Two tennis balls bouncing around in a large room is just pure fantasy. The reporter hasn't done the math. The reporter doesn't really understand what 0.0000336% really means.

Fred Gauss

This is not Fred.

If Fred were wearing a fake mustache and a top hat to disguise his square head, there is still an easy way to spot Fred.

his eyes!

TWO TENNIS BALLS IN A LARGE ROOM?

diameter of a tennis ball	2.6 inches
radius of a tennis ball	1.3 inches
volume $(4/3)\pi r^3$	11.9636 cubic inches

A standard percent problem: *11.9636 cubic inches is 0.0000336% of what?* You divide the number closest to the *of* into the other number.

$\frac{11.9636}{3.36 \times 10^{-7}}$ 3.56059 × 10^7 cubic inches

Convert to cubic feet. The conversion factor is $\frac{1 \text{ cubic foot}}{12^3 \text{ cubic inches}}$ 123,631 cubic feet

If a house had nine-foot ceilings, its square footage would be 13,736 square feet

Two gold atoms the size of tennis balls would be bouncing around in 27,472 square feet of house.

If you take your standard six-bedroom, four-bath mansion, it would take about five of these mansions to equal the space in which those two tennis balls were trying to find each other.

That is very different than six people squeezed onto a park bench.

That is why it does not make much difference whether the gas particles (atoms or molecules) are 1 Å or 5 Å.

Your Turn to Play

1. Three pages ago when Fred was doing the arithmetic, he left out some of the steps.

For example, he went from the radius of a gold atom 1.44 Å to its volume of 1.25×10^{-29} L.

Please fill in the steps.

2. On his next step he went from the volume of one gold atom, which is 1.25×10^{-29} L, to the volume of a mole of them, which is 7.53×10^{-6} L.

Please fill in the steps.

3. How many atoms of N_2 (nitrogen gas) are in a 4.00 L container that is at 30.0° C at a pressure of 400 mm of mercury?

(This is a big conversion factor problem. Start with one mole of N_2 at STP occupying 22.4 L. ← That's the one chem fact we use over and over again.)

It might take you 15 minutes to solve this third problem. That is okay. Some readers may finish it in 5 minutes. Some, in 30. This is not a horse race.

The answer will be 5.10×10^{23} atoms. If you get an answer of 5.09×10^{23} or 5.11×10^{23}, that is just fine. That proves you have done the right steps. Calculators can give slightly different answers.

.......COMPLETE SOLUTIONS.......

1. Volume of a sphere $= (4/3)\pi r^3 = (4/3)\pi(1.44\ \text{Å})^3$
$= (4/3)(3.14159)(1.44 \times 10^{-10}\ \text{m})^3$
$= (4/3)(3.14159)(2.985984)(10^{-30}\ \text{m}^3)$
$= 12.50764996608 \times 10^{-30}\ \text{m}^3$
$= 1.25 \times 10^{-29}\ \text{L}$

2. $\dfrac{1.25 \times 10^{-29}\ \text{L}}{1} \times 6.02 \times 10^{23}$ in one mole $\doteq 7.53 \times 10^{-6}$ L in one mole

3. $\dfrac{\text{1 mole of } N_2 \text{ at STP occupying 22.4 L}}{1} \times$

$$\frac{4.00\ \text{L}}{22.4\ \text{L}} \times \frac{400\ \text{mm of Hg}}{760\ \text{mm of Hg}} \times \frac{273\ \text{K}}{303\ \text{K}} \times \frac{6.02 \times 10^{23}\ \text{atoms}}{1\ \text{mole}}$$

$= 5.10 \times 10^{23}$ atoms

The conversion factor that most often gives difficulty in this problem is the $\dfrac{273\ \text{K}}{303\ \text{K}}$

Here's the reasoning. Suppose the volume is constant. Suppose the pressure is constant. Suppose I'm heating up the container.

The only way I can keep the pressure constant is let some of the molecules out of the container. If I didn't, the pressure would go up.

So as I heat up the container, I'll expect fewer atoms. That's why the smaller number $\dfrac{273\ \text{K}}{303\ \text{K}}$ is on top.

Chapter Nine
Two Worlds

Fred told his class that the everyday things of baking cookies,
burning cookies,
digesting cookies, and
using soap to wash up
are all chemical processes. Even though those processes are happening down at the atomic level, they *feel real* to us. When cookies are baking, they smell real. Tasting warm chocolate chip cookies is certainly real.

On the other hand, the stuff that happens down at the atomic level is like visiting another planet . . . or another universe. It is a bit like Alice's Adventures in Wonderland, except that the white rabbit doesn't speak English. He speaks Finnish, Greek, or Arabic (take your choice).

For example, it really doesn't seem real that a balloon filled with any gas has atoms or molecules that are so far apart from each other that they would be like two tennis balls bouncing around in five mansions.*

We don't live and think at the atomic level. And because of that, we need to build bridges between the atomic world and ours.

We have already build a "distance bridge." It's much nicer to say that the radius of an atom of gold is 1.44 Å, than to write that the radius is equal to 0.000000000144 meters. The diameters of atoms are generally between one and five angstroms. That sounds friendly.

---

* The big idea is what is important: *gas molecules are far apart.* Air in a balloon isn't at STP. The pressure higher than 760 mm of mercury because of its confinement in the balloon. (That would tend to add more molecules.) But it is also at a higher temperature than 273 K. (That would tend to have less molecules.) Stay loose. ☺ The big picture is how incredibly far apart those molecules are from each other.

Let's not be like that famous group of people who strain at gnats.

We have built a "number bridge." We talk about a mole of particles (also known as Avogadro's number). That is much easier on the pencil than writing 602,000,000,000,000,000,000,000.

We are about to build a "weight bridge". Instead of weighing individual helium atoms (0.000 000 000 000 000 000 006 66 grams), we will work in moles of atoms. Things get a lot nicer. A mole of helium weighs 4.003 grams. A mole of oxygen atoms weighs 16.00 grams. (You computed these in the *Your Turn to Play* on page 46.)

The **atomic weight** of helium is 4.003.

The atomic weight of oxygen is 16.00.

The atomic weight of hydrogen is 1.008.

A molecule of water (H_2O) consists of two atoms of hydrogen and one atom of oxygen.

The **molecular weight** of water is 18.016. (The math: 16.00 + 1.008 + 1.008)

The molecular weight of a molecule of oxygen (O_2) is 32.00. (The math: 16.00 + 16.00) Give me 32 grams of oxygen and I know that I have a mole of oxygen molecules and that at STP it will occupy 22.4 liters.

Things are so much happier when we are working in moles rather than down in the tiny world of individual atoms.

Ten notes

♪#1: Helium is not very friendly. It likes to go around as a single atom. It never pairs up with anything else. In contrast, oxygen almost always likes to pair up as an O_2 molecule. If you put a bunch of single atoms of oxygen in a dance hall, O O O O O O O O O O O O O O O O O O O O O O O O O O O O O O O O it wouldn't be long before they were all dancing in pairs O_2 O_2 O_2 O_2 O_2 O_2 O_2 O_2 O_2 O_2 O_2 O_2 O_2. The oxygen you breathe in is almost all O_2.*

---

* If you put electric sparks through a bunch of dancing O_2 molecules you can mess things up. Three O_2 molecules will turn into two O_3 molecules of **ozone**, which is a pale-blue poisonous gas. $3\,O_2 \rightarrow 2\,O_3$. Ozone molecules quickly realize their mistake and decompose back into O_2.

♪#2: Nitrogen, chlorine, bromine, and hydrogen all like to dance in pairs. They naturally form **diatomic molecules**: N_2, Cl_2, Br_2, and H_2.

♪#3: If helium is unfriendly (and doesn't form compounds), and if nitrogen, chlorine, bromine, and hydrogen like to go around in pairs, then the carbon atom is—how shall we say it?—nuts. It's crazy. It likes to combine in so many different ways that it makes almost every other element boring. Whole chemistry books and chemistry courses are devoted to just the compounds that involve carbon. The chemistry of carbon is called **organic chemistry** (organic = living).

The border between the fields of chemistry and biology is organic chemistry.

Hydrogen, sulfur, and oxygen might form sulfuric acid (H_2SO_4), which has an molecular weight of 98. (The math: $2\times1 + 32 + 4\times16 = 98$) A mole of sulfuric acid would weigh 98 grams.

In contrast, an organic (carbon-containing) molecule, such as DNA which is found in cells in your body, can have a molecular weight of 16,000,000.

♪#4: If you know that *the weight (in grams) of a mole of atoms of any particular element is its atomic weight,* then you probably know more about atomic weight than most adults at the grocery store.*

And *the molecular weight of any molecule is the weight (in grams) of a mole of those molecules.*

And the weight of a dozen pounds of ground beef is 12 pounds. (I'm being silly.)

♪#5: Now, a lesson in polite language. Around your close friends after you have had a couple of milkshakes, you might use "words" like *ain't* or *alright*** or *irregardless*, even though everyone knows that those are words used by the uneducated.

---

* If you don't believe me, walk up to someone and ask, "Pardon me. I've got a couple of questions. Do you know where the carrots are and what does it mean when we say that the atomic weight of helium is 4.003?"

Most people will be able to answer 50% of those questions right.

** Because *already*↔*all ready* form a nice pair, some assume that there is a similar pairing with *all right*. There isn't. Note that *already* and *all ready* have different meanings. Compare "Are we there already?" with "Is everyone all ready to go?"

Similarly, chemists, when they are around friends, talk about atomic *weight*. Everyone knows that's improper. Grams are a measure of mass, not of weight. (While no one is watching, you can sneak a look back to the essay on Mass vs. Weight on page 35. I won't tell anyone.)

When those chemists give speeches in front of the Super Professional Elite Chemists of the World Society, they never say *atomic weight,* but say **atomic mass**. It means the same thing. It's just ~~correcter~~ more correct.

♪#6: It even gets ~~worser~~ ~~more worse~~ worse. When chemists give acceptance speeches when they are receiving Nobel prizes in chemistry, they don't say *atomic mass*—that would be so sloppy. They say **average atomic mass**. This is the most accurate, most proper, expression. When you are wearing a tuxedo, you don't say *a mole of stuff in grams.*

Guide to Correct Usage

formal	business	relaxed
average atomic mass	*atomic mass*	*atomic weight*

Questions from readers . . .

Why say average atomic mass? You weigh a mole of helium (22.4 liters at STP) and you have the atomic weight of helium.

Does average mean the average of a lot of weighings? No.

Do different helium atoms have different weights? Forget that question! That would be silly. Atoms are the basic building blocks of the universe. Yes. ***And all helium atoms would then have to be alike.*** No.

Wait! Stop! What do you mean by " no ***"? I, your reader, know a little about chem. Aren't all protons exactly identical?*** Yes. ***And aren't all neutrons exactly identical?*** Yes. ***And aren't all electrons exactly identical?*** Yes. ***Aren't all helium atoms (at least as far as high school chemistry is concerned) all made up of protons, neutrons, and electrons?*** Yes.

Then, by logic, wouldn't all helium atoms have to be identical? No.

Mr. Author, you are being difficult and ornery. I don't have room to explain here. Let me explain in Chapter 10.

Your Turn to Play

1. We have mentioned several molecules that consist of two atoms (diatomic molecules). N_2, Cl_2, Br_2, and H_2.

Let's test your memory. We have mentioned two molecules that consist of three atoms. Name them.

2. Isn't it nice that water (H_2O) and air (N_2 and O_2) have no color and no odor and that poisonous stuff like chlorine gas (Cl_2), bromine (Br_2), and ozone (O_3) all stink?

Multiple choice answer: □ yes or □ yes. (Take a wild guess.)

3. A little more practice with conversion factors. How many molecules of hydrogen (H_2) are in a 10.00-L container at STP?

4. The chemical symbol for the element copper is Cu. How many atoms of copper are in one mole of copper in a 10.00-L container at STP?

5. In everyday life density is something you run into much more frequently than atomic weight. In Chapter 5, we mentioned three times that 1 mL of water weighs 1 gram. The density of water is 1g/mL. "One gram per milliliter." This could also be written as $1g/cm^3$ or as 1g/cc.

The **density** is measured in mass per volume.

The atomic weight of Cu is 63.55. A solid block of copper containing a mole of copper atoms is dropped into a full glass of water. It sinks to the bottom. 7.12 grams of water overflows onto the table. What is the density of copper?

.......COMPLETE SOLUTIONS.......

1. Water (H_2O) and ozone (O_3).

There are zillions of other molecules with three atoms. If you drink bubbly soda, the bubbles are carbon dioxide (CO_2).

Logic lesson: Every organic compound contains carbon atoms.

This does not imply that CO_2 is an organic compound.

2. I think it's nice. ⊠ yes

3. We start with what we know. There is 1 mole of hydrogen in 22.4 L at STP.

$$\frac{\text{1 mole in 22.4 L}}{1} \times \frac{\text{10.00 L}}{\text{22.4 L}} = \text{0.446428 moles in 10.00 liters}$$

$$\frac{\text{0.446428 moles in 10.00 L}}{1} \times \frac{6.02 \times 10^{23}\text{ molecules}}{\text{1 mole}}$$

$= 6.02 \times 10^{22}$ atoms in 10 L.

4. Do you think that pennies are a gas at standard temperature of 0° C? Drop a mole of copper into a 10-liter container and it goes "clunk." A mole of *anything* is 6.02×10^{23}.

5. What's the mass of a mole of copper, which has an atomic weight of 63.55? It's 63.55 grams.

What's the volume of 7.12 grams of water? It's 7.12 mL.

What's the volume of the block of copper that displaces 7.12 mL of water? It's 7.12 mL.

$$\text{So the density} = \frac{\text{63.55 grams}}{\text{7.12 mL}} \doteq \text{8.93 grams per cm}^3$$

Copper is almost nine times denser than water.

Chapter Ten

All Helium Atoms Are Not Created Equal

Stop! Before you start this chapter, Mr. Author, you need to explain one thing.

What's that?

In #4 of the *Your Turn to Play* you slipped in the fact that the chemical symbol for copper is Cu. That's crazy. Why isn't it Co?

That's easy. Co is the symbol for cobalt, which is a different chemical element.

But why Cu? There isn't any "u" in the word copper.

Cu is from the Latin word *cuprum,* and *cuprum* comes from the Latin *Cyprium*, which means from Cyprus, which was an important ancient source of the metal copper.

Cyprus? Isn't that a tree?

You are thinking of cypress trees. Cyprus is an island.

Fred gave his usual *History of the Atom* lecture at this point. He started with the Greek guy named Democritus in 430 BC and went to 1932 when James Chadwick discovered the neutron.*

---

* It's all stuff you read in *Life of Fred: Elementary Physics* on pages 229–244 when you were in middle school years. I'm not going to repeat those 16 pages here. ("16 pages" is not a typo.)

The important part of Fred's History of the Atom lecture was his description of protons, neutrons, and electrons in an atom. There is a bunch of other stuff—particles and forces—in atoms, but those are of interest mostly to physicists. You will be able to do chemistry for a long, long time just dealing with protons, neutrons, and electrons.

Let's start with a Basic Picture of an atom.

In the middle (the nucleus) are protons and neutrons. Flying around the nucleus are the tiny electrons.

The protons and neutrons are like people huddling together in fear of the vultures (electrons) that are circling overhead. The electrons are sometimes in orbits farther away from the nucleus, and often they drop into lower orbits. But they don't fall into the nucleus. (I don't know why.)

The electrons have a negative charge. The protons have a positive charge. And the neutrons are neutral—no charge at all.

The protons and neutrons are big and fat. A proton weighs 1.0073 atomic mass units (amu) and a neutron weighs a bit more: 1.0087 amu. An electron is a skinny 0.0005486 amu. (We'll talk about atomic mass units in a moment.)

Wait a minute! I, your reader, have a question. How much electrical charge is on an electron?

–1 (and some chem books write that as 1–)

What's the electrical charge on a proton?

+1 (and some chem books write that as 1+)

But a proton is about 1836 times larger than an electron! (I divided 1.0073 amu by 0.0005486 amu.) That doesn't make sense. I'll draw it for you. Nature couldn't have made it that way.

Just because it doesn't seem reasonable to you is not a great argument about what is true. Do you remember when you first learned about how babies are made? Your first response was, "My parents would never do *that*!"

When we talked about the scientific method (back on page 47), there were three paths to the truth: ask yourself, ask an authority, and ask nature. When we do experiments—when we ask nature—we find that protons are about 1836 times heavier than electrons, even though their electrical charges are equal (and opposite).

I agree. It really doesn't make any sense at all. If I had created the universe, I would have made subatomic* particles with a positive charge exactly the same size as subatomic particles with a negative charge. That would make sense.

I have decided to accept Reality:

1) I didn't create the universe.

2) Protons are much bigger than electrons.

And now to answer your question from the previous chapter about all helium atoms not being identical and the question of what an atomic mass unit (amu) means, let's go back to listening to Fred's lecture about atoms.

Fred continued, "Hydrogen atoms each contain one proton. Helium atoms each contain two protons. Carbon has six protons. Nitrogen has seven. Oxygen has eight."

One of the reporters raised his hand. "Do you mean that there aren't any atoms with three or four or five protons?"

Fred pointed to the **Periodic Table of the Elements** poster that was on the side of the lab table.

1 H 1.00794							2 He 4.00260
3 Li 6.941	4 Be 9.01218	5 B 10.811	6 C 12.0107	7 N 14.0067	8 O 15.9994	9 F 18.9984	10 Ne 20.1797
11 Na	12 Mg	13 Al	14 Si	15 P	16 S	17 Cl	18 Ar

* Protons, neutrons, and electrons are called subatomic particles because they are all part of an atom. That *does* make sense.

Fred looked at the poster. The bottom had been torn off when the lab table had been pushed from the chemistry classroom to Fred's classroom.

Fred answered the reporter. "If an atom has three protons, it is Li." He wasn't sure what Li stood for. Bob shouted from the back of the room, "Li is lithium. Be is beryllium. And B is boron."

Fred liked to talk about hydrogen, helium, and carbon because those were elements that many people have heard about. Who has ever heard of beryllium?

Fred explained, "The numbers on the periodic table above the symbols indicate the number of protons. Helium has two protons. Instead of saying the long phrase, "Helium has two protons," chemists like to shorten things up by saying, 'The **atomic number** of helium is two.'"

Fred realized how stupid that sounded. He blushed.

"Or maybe they shorten it to 'Helium's atomic number is two.'"

He carefully wrote out both sentences on the board:

Helium has two protons.

Helium's atomic number is two.

The atomic number sentence was still longer. He erased and tried again:

If you count the number of protons in a helium atom, it equals two.

Helium's atomic number is two.

Fred unblushed. He had "proved" that using the words *atomic number* made things shorter.

No one was convinced.

Fred explained the rest of the periodic table. "The number under the symbol is the average atomic masses, which are measured in atomic mass units (amu). The average atomic mass of helium is 4.00260.

Wait! I, your reader, must object. You said that in this chapter you would tell me why all helium atoms are not alike and what amu means.

I'm outta room. I'll try to get to it in Chapter 11. Sorry. ☺ We have to do Your Turn to Play.

Your Turn to Play

1. Back at the end of Chapter 6 in problem #1 (page 46) we took the weight of a single atom (0.66495×10^{-23} grams) and used conversion factors to find the weight of a mole of helium.

The computation looked like this:

$$\frac{1 \text{ mole of He}}{1} \times \frac{6.02 \times 10^{23} \text{ atoms}}{1 \text{ mole}} \times \frac{0.66495 \times 10^{-23} \text{ grams}}{1 \text{ atom of He}}$$

$\doteq$ 4.003 grams

Looking at the periodic table (two pages ago), we read that the atomic weight is 4.00260. Make a guess why these two numbers are different.

2. ✓ A mole of protons weigh about 1 gram.

✓ A proton and a neutron weigh about the same. (If you want to be fancy, you would say that they have roughly the same mass.)

✓ 22.4 liters of helium at STP weighs about 4.003 grams.

From this information, determine how many neutrons are in an atom of helium.

3. Two liters of chlorine gas (Cl_2) at 10° C and at a pressure of 1.2 atmospheres weighs 3.664 grams. What is the atomic weight of chlorine?

(Start with what you know: 3.664 grams. Then use conversion factors.)

4. Uranium (chemical symbol U) is the main "ingredient" in making atomic bombs. It is metallic and silvery-white and has 92 protons. Eight years before uranium was discovered, the planet Uranus was discovered. Uranium was named after the planet. What is uranium's atomic number?

5. Neptunium has atomic number 93. Make a guess what neptunium was named after.

.......COMPLETE SOLUTIONS.......

1. When you round off 4.00260 to four significant digits you get 4.003, which was the number that Fred arrived at using some instruments at KITTENS University.

The 4.00260 in the periodic table was found using much better equipment than was available to Fred.

2. The atomic number of helium (from the periodic table) is 2. There are two protons in each atom of helium.

In 22.4 liters of helium at STP there is one mole of helium atoms. There are two moles of protons in the 22.4 liters.

Since each mole of protons weighs about one gram, the protons account for two grams of the 4.003 grams.

The other two grams come from the neutrons. There must be two moles of neutrons in the 22.4 liters.

Each atom of helium must have two neutrons.

3. 3.664 grams in 2 L at 10° C at 1.2 atm.

$$\frac{3.664 \text{ grams}}{1} \times \frac{22.4 \text{ L}}{2 \text{ L}} \times \frac{283 \text{ K}}{273 \text{ K}} \times \frac{1 \text{ atm}}{1.2 \text{ atm}} = 35.4499 \doteq 35.45$$

You could argue that the 35.45 should have been rounded off to one or two significant digits.

4. Uranium has 92 protons. Its atomic number is 92.

5. Neptunium was named after the planet Neptune.
And plutonium (Pu), which was the main "ingredient" in the second atomic bomb in WWII (over Nagasaki), was named after Pluto.

Chapter Eleven
Etymology

Fred needed to explain why chemists say "average atomic mass" rather than just "atomic mass." Since all protons are alike and all neutrons are alike, you might think that every helium atom with two neutrons would all have the same mass.

By definition, a helium atom has two protons. That's its atomic number. The **mass number** of an atom is the number of protons plus the number of neutrons. A helium atom with two protons and two neutrons has a mass number of exactly four.

In chemistry the mass number is written in a weird position. A helium atom with a mass number of 4 is written as ^{4}He.

Okay. Okay. Okay. I, your reader, am getting a little antsy. I've been waiting since the end of Chapter 9, to get my questions answered. Instead, you've been fooling around drawing a picture of the island of Cyprus. Answer me now. Do all the ^{4}He atoms weigh the same?

Yes.

So why in blazes do you talk about average atomic masses?

Because there are other helium atoms. There are helium atoms with two protons and no neutrons: ^{2}He.

Two protons and one neutron: ^{3}He.

The most common helium atom: ^{4}He.

And really rare ones: ^{5}He up through ^{10}He.

There are nine **isotopes** of helium. Isotopes of any element all contain the same number of protons but different numbers of neutrons.

So when you weigh 22.4 L of helium at standard temperature and pressure, you are weighing more than one variety of helium atoms. That is why chemists talk about *average* atomic mass.

Great. You got that settled. Now, before you get distracted by stuff like Uranium was named after the planet Uranus, tell me how atomic mass units (amu) fit in with the 22.4 L at STP stuff.

But the etymology (the history of words) is so much fun. I like it better than doing conversion factor problems. It was a delight to me when I found out that the element polonium was named after the country Poland and terbium was named after Ytterby, which is a town in Sweden.

The chemical element nickel has the most interesting etymology. Type in `etymology nickel` into any search engine on your computer. Some of the articles will trace *nickel* back to Satan. We have an element named after the devil.

Can you tell that I really don't want to talk about atomic mass units?

Yup. It's pretty obvious.

Okay. Here goes. **One amu is defined** to be exactly one-twelfth ($\frac{1}{12}$) of the mass of ^{12}C. Since the atomic number (the number of protons) in carbon is 6, one amu is one-twelfth of six protons and six neutrons.

One amu equals 1.66054×10^{-24} grams.

The average atomic masses listed in the periodic table are given in atomic mass units—virtually, but not quite, the weight of a mole of atoms measured in grams.

The League of Chemists (or whatever it's called) made the switch from weight of a mole of atoms measured in grams to amu to make everything more ~~exacter~~ exact. It is like when physicists changed the definition of a meter from the distance between two scratches on a metal bar (as defined in 1889) to $\frac{1}{299{,}792{,}458}$ of the distance that light can travel in one second in a vacuum, which became the new definition in 1983.

For you and for me, it really doesn't make much difference. And to make everything just a bit more difficult, the International System of Units (known as SI) has declared that amu is to be written as u. (Lots of people still write amu.)

I NEED TO TAKE A BREAK

I got up the courage to talk about average atomic masses and atomic mass units—things I've been putting off for over a chapter. Those topics are just a bit too fancy for my taste. I'm more of an atomic weight guy than an average atomic mass guy. I'd rather measure a hundred meter track with a meter stick (which is like a 39-inch yardstick) than suck all the air out (to make a vacuum) and then time how long light took to "run" that distance and then multiply by 299,792,458.

Stan's Playtime

No other chem book talks about the nine isotopes of helium. Instead, lots of them like to mention the first three isotopes of hydrogen: ^{1}H, ^{2}H, and ^{3}H. (And they don't mention ^{4}H, ^{5}H, ^{6}H, or ^{7}H at all.)

^{1}H has a mass number of 1. It consists of a single proton (and a single electron flying around it). In any atom the number of electrons always equals the number of protons. That makes the atom electrically neutral.

^{1}H is pretty popular (= 99.98% of all hydrogen atoms). Its nucleus, a single proton, is considered fairly stable. No one has ever seen a proton fall apart. (The technical word is *decay.*) Scientists estimate that there is a half chance a proton will decay in 10^{36} years. (1,000,000,000,000,000, 000,000,000,000,000,000,000 years) How they figured that out, I'll never know. That's why they say that ^{1}H is fairly stable. (Actually, they say *very* sable.)

^{2}H consists of a proton and a neutron. I like to think of a neutron as FAT. So ^{2}H is half fat and half muscle. It is also a very happy (and stable) atom like its anorexic sister ^{1}H.

^{3}H has one proton and two neutrons. It's two-thirds FAT. It's life span is a lot shorter than ^{1}H or ^{2}H. If you have a kilogram (or a pound or a ton) of ^{3}H, half of it will decay in 12.32 years. And half of what's left will decay in another 12.32 years. I wouldn't sell life insurance to someone as fat as ^{3}H. Atoms whose nuclei (plural of *nucleus*) are unstable (= fall apart or decay) are called **radioactive**.

None of the heavier isotopes (^{4}H, ^{5}H, ^{6}H, and ^{7}H) can be found in nature. They have been created in labs. All of them are so fat that their half-lives are each less than a zeptosecond.*

---

* On the *Ye Old Metric Prefixes* poster in Chapter 3 were centi = hundredth and milli = thousandth. It didn't mention the less-common zepto = 10^{-21}, which is one-thousandth of a millionth of a millionth of a millionth.

Hey! You aren't going to leave me with just centi, milli, and zepto, are you?

This was supposed to be *my* playtime. I'm the author. Don't I get a little playtime?

I bought your book. It's my playtime now. Gimme all those metric prefixes . . . please.

Okay. But I bet you a megabuck (a million dollars) you'll never use most of them.

yotta	10^{24}	
zetta	10^{21}	
exa	10^{18}	
peta	10^{15}	
tera	10^{12}	Some hard drives are rated in terabytes.
giga	10^{9}	
mega	10^{6}	
kilo	10^{3}	A kilometer is a thousand meters.
hecto	100	
deca	10	
deci	10^{-1}	
centi	10^{-2}	A centidollar is a penny.
milli	10^{-3}	A milliliter is a thousandth of a liter.
micro	10^{-6}	
nano	10^{-9}	
pico	10^{-12}	
femto	10^{-15}	
atto	10^{-18}	
zepto	10^{-21}	
yocto	10^{-24}	

$\approx$ means "approximately equal to"

One lifetime $\approx$ 85 years

$$= \frac{85 \text{ years}}{1} \times \frac{365\tfrac{1}{4} \text{ days}}{1 \text{ year}} \times \frac{24 \text{ hours}}{1 \text{ day}} \times \frac{3600 \text{ seconds}}{1 \text{ hour}}$$

$$\doteq 2{,}682{,}396{,}000 \text{ seconds} \doteq 2.7 \text{ gigaseconds}$$

Your Turn to Play

1. The atomic number of carbon (C) is 6. How many neutrons are in the isotope ^{14}C?

2. When you are not doing hard work such as *Your Turn to Play* problems, your respiration rate may be about 12/minute.

That translates into one inhalation and exhalation every 5 seconds.

Using a conversion factor, convert 5 seconds into nanoseconds.

3. Just for fun, let's go back about a hundred years ago (roughly speaking) when you learned to do percent problems. When you are doing *Your Turn to Play* problems your respiration rate is probably 60% higher than 12/minute. How fast is that? (If you get an answer that is less than 12, you know that it isn't correct.)

4. [easy problem] The half-life of ^{3}H is 12.32 years. If you start with 4 grams of it, how long will it be before it is only 1 gram?

5. [hard problem that can be skipped unless you have had two years of high school algebra] How long will it take for 4 grams of ^{3}H to become 0.55 grams?

Here's a start on that problem.

4 grams becomes $4(\frac{1}{2})$ grams after 12.32 years.
4 grams becomes $4(\frac{1}{2})^2$ grams after 2(12.32) years.
4 grams becomes $4(\frac{1}{2})^3$ grams after 3(12.32) years.
4 grams becomes $4(\frac{1}{2})^x$ grams after x(12.32) years.

We want x so that $4(\frac{1}{2})^x = 0.55$

After we find x, then we know that in x(12.32) years the 4 grams will turn into 0.55 grams.

.......COMPLETE SOLUTIONS.......

1. ^{14}C has a mass number of 14. Mass number = protons + neutrons.
   C has an atomic number of 6. Atomic number = protons.
   ^{14}C has 8 neutrons.

2. $\dfrac{5 \text{ seconds}}{1} \times \dfrac{1 \text{ nanosecond}}{10^{-9} \text{ seconds}} = 5 \times 10^9$ nanoseconds
   $= 5{,}000{,}000{,}000$ nanoseconds

3. In *Life of Fred: Decimals & Percents* you learned two ways to do "60% More" problems.

THE HARD WAY: First find 60% of 12. $0.6 \times 12 = 7.2$
Second, add that 7.2 to the original 12. $7.2 + 12 = 19.2$

THE EASY WAY: A 60% gain means the original 12 (100%) plus an extra 60%. 100% + 60% = 160%.
$12 \times 1.6 = 19.2$

4. 24.64 years
After 12.32 years, the 4 grams will become 2 grams.
After another 12.32 years, the 2 grams will become 1 gram.

5. $4(\frac{1}{2})^x = 0.55$ is an exponential equation. The unknown is in the exponent. We solve $4(0.5)^x = 0.55$ using logs.

Divide both sides by 4	$(0.5)^x = 0.1375$
Take the log of both sides	$\log (0.5)^x = \log 0.1375$
$\log a^x = x \log a$ (the Birdie rule)	$x\log (0.5) = \log 0.1375$
Approximate the logs with a calculator	$x(-0.30103) = -0.8617$
Algebra	$x \doteq 2.8625$ half-lives

$$\frac{2.8625 \text{ half-lives}}{1} \times \frac{12.32 \text{ years}}{1 \text{ half-life}} \doteq 35.27 \text{ years}$$

(We rounded to four significant digits because 12.32 has four significant digits.)

Chapter Twelve

Mixtures

Fred needed to do some "real" chemistry. When most students think of chemistry, they think of test tubes and Bunsen burners, not of average atomic masses. They want to see stuff happen.

"Chemistry is a little like English," he lectured. "Writers start with the 26 letters of the alphabet."

He revised that. "They start with the 52 lower case and capital letters of the alphabet."

That still wasn't quite right. "They start with 52 letters and punctuation and special symbols."

He wrote on the board: abcdefghijklmnopqrstuvwxyzABCDEFGHIJ KLMNOPQRSTUVWXYZ.,;:' ' " "?!/@#$%^&*()[ ]{ } - + – — 0123456789

"And from these roughly 90 symbols, all of the stories, poems, and love letters have been written.* The same is true in chemistry. We start with 90 naturally occurring elements—some books say 88—and from them come cherry pies and concrete and sodium chloride (table salt, NaCl).

"Some are **compounds**, (definition: molecules of at least two different elements) such as NaCl, H_2O, CO_2, H_2SO_4. In compounds, the proportions are fixed. In sulfuric acid (H_2SO_4) there are exactly two atoms of hydrogen, one atom of sulfur and four atoms of oxygen that combine together to make a molecule of sulfuric acid.

"Others are **mixtures**, such as when you make mud pies. You can use a lot of water with a little dirt, or you can use a little water with a lot of dirt." Joe got excited. He knew that if they were going to make mud pies in chemistry, he would be the expert.

Fred wrote on the board: Compounds—fixed proportions. Mixtures—variable amounts.

---

* If Fred had been lecturing in Greece (AαBβΓγΔδEεZζHηΘθιKΛλM μNνΞξOoΠπPρΣσΣςTτYυΦφXχΨψΩω) or in Russia (AaБбBвГгДдEe ЁёЖжЗзИиЙйКкЛлМмНнОоПРрсТтуФфхЦЧчШшЩщъЫыЬьЭэЮю ЯяӘәЃѓҒғЂђЄє), things would be slightly different.

Joe put up his hand. "Could we start with mixtures? I like them better." What Joe was thinking was that there would be less math involved—no fixed proportions.

Fred asked, "Would you want to assist me, Joe, in our first mixture experiment?"

"Sure. I'd be glad to." He stood up. The jelly beans on his lap spilled to the floor.

"Here are two flasks," Fred told Joe.

Joe counted them and nodded.

"Carefully pour them into a big bowl and stir them."

Joe did it.

Fred thanked him, and Joe went back to his seat. Joe decided that he was going to phone his mother this afternoon and tell her about the Great Chemistry Experiment that he had participated in.

"This is a mixture," Fred said. "Since I can still see the various particles of sugar and sand, this is called a **heterogeneous** (het-er-eh-GENIUS) mixture.

"Can anyone think of a way to separate the sugar from the sand?"

"It's going to take a long time," Joe said, "but I can do it." He held up a pair of tweezers.

Betty, one of the brighter chem students, raised her hand. "Add some water. The sugar will dissolve. Then pour off the water. Do this a couple of times and just the sand will remain."

"Perfect," said Fred.

Readers who have read other Life of Fred books know that Joe's girlfriend, Darlene, has been trying to get Joe to marry her for a long time. For their honeymoon Darlene has shown Joe pictures of the white beaches of Hawaii. Joe always thought that those beaches were sugar. Darlene did not disabuse him of that thought.

"Dust floating in the air is another heterogeneous mixture. Can anyone think of a way to separate that mixture?"

Lots of hands shot up. Air filters in furnaces and in vacuum cleaners are designed to do that. Coffee filters do that for the coffee grounds. Cars have gasoline filters that filter out any solid particles before they can mess up the engine. Cigarette filters remove some of the crud from the smoke, but smoking is still like playing dodge ball in front of a firing squad.

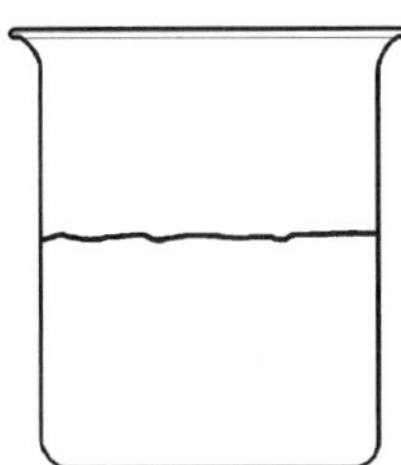

"Here is a beaker of salt water. You can't see particles of salt in the water. This is called a **homogeneous** mixture. If you want to separate the salt from the water, tweezers won't work. Coffee filters won't work. Does anyone have an idea how to separate this mixture?"

Betty suggested letting the water evaporate, which would leave just the salt. Bob Bunsen suggested boiling the water off using, of course, a Bunsen burner.

Fred continued, "Gold, which has the chemical symbol Au, can be found. . . ."

Darlene, who always sits next to Joe in every class he takes, shouted, ". . . in wedding rings!" She held up three wedding ring brochures that she had been showing to Joe.

As a good teacher, Fred knew that often students would learn more from someone who was really excited about a subject. Darlene was obviously excited. Fred asked her to step up in front and show the class.

"When I was first looking at wedding rings," she began, "I wanted solid gold rings for Joe and me. I thought that would look nicest next to my white wedding dress and his black tuxedo, but none of the jewelers would sell me pure gold rings.

"When I asked them, they said that pure gold—which they called 24 karat gold—is way too soft and bendable."

Fred wrote on the board: 24 karats = 24K = 24 kt

24 karats = pure gold.

18K = 18/24 gold + 6/24 other metal(s)

Darlene continued, "So I asked how we could toughen up the gold. They told me that when you alloy gold with other metals, it gets harder and more durable. She read from one of the brochures: red gold = gold + copper. rose gold = gold + copper + silver. white gold = gold + platinum. gray-white gold = gold + iron + copper. soft green gold = gold + silver. purple gold = gold + aluminum.

"I'm going to pick which color alloy depending on which color frosting will be on our wedding cake."

Darlene sat down. Joe was reading a fishing magazine. He wasn't interested in wedding rings and cakes.

Fred explained alloys. "To make an alloy of gold and copper [red gold] you take gold and copper and mix them together."

The whole class gave Fred blank looks. That really didn't make much sense. They imagined tossing a bar of gold and a bar of copper into a bowl and mixing them.

"Of course," Fred said, "you have to melt them first. I can't do that very easily with the chem equipment on this lab table. The best I can do is melt some glass tubing. So gold alloys are gold mixtures.

"But gold isn't the only metal mixture. Imagine you had a sword made out of pure copper—like the copper in copper electric wires. One good whack and the stupid sword would bend. About 4,000 years ago, copper and tin were alloyed to make bronze, which doesn't bend as easily as copper. The armies that had bronze swords had a big advantage over the copper sword armies. It was the beginning of the Bronze Age.

"Brass came later—an alloy of copper and zinc. It reflects sound well and is resistant to salt water (and spit). Trombones, trumpets, and tubas are in the brass section of an orchestra.

"Perhaps more important than jewelry, swords, or tubas is steel. Pure iron [chemical symbol Fe] is flexible and malleable.* Alloy iron with a bit of carbon and you get steel. Add a fifth of one percent (0.2%) carbon to iron and you can make nails and chains. Add one percent carbon to iron and it gets really hard—razors, knives, and drills."

Fe (iron) comes from the Latin *ferrum.* That's kind of boring. My favorite is Hg (mercury) which comes from the Latin *hydrargyrum,* which means water silver. The prefix *hydra* is part of our words hydrant and hydrate.

Of the chemical elements that are metals, only Hg is liquid at room temperature.

---

* malleable = can pound it flat with a hammer and it stays in one piece. Sugar cubes are *not* malleable.

Your Turn to Play

1. Iron is softer than steel. Files are made of steel. So when you rub a file on a block of iron, you get iron filings, not steel filings.

Take the iron filings and mix them with sand. Is the mixture heterogeneous or homogeneous?

Please remember to write out your answers for each of the questions in the Your Turn to Play. You will learn more than just saying the answers. And, in the case of this question, you will also start to learn to spell a six-syllable word with four e's in it.

2. How could you separate iron filings from sand?

3. If you put some food coloring (or some ink) into water and stir it, you get a nice homogeneous (only five syllables and two e's) mixture.

If you stir oil into water and wait a couple of seconds, the oil floats to the top. Then you can pour off the top layer. In chemistry that is called **decanting**.

not homogeneous

Name one other way—other than decanting—to separate this mixture. (I can think of three ways.)

4. If you stick your greasy, grimy, oily hands under cold running water, they stay greasy, grimy, and oily. In this situation you can't decant off the oil. What do you do?

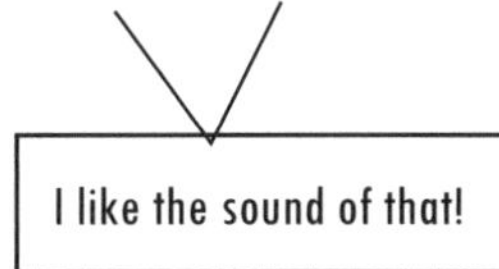

5. The recipe for Joe's famous sky-blue mud pie: 3 handfuls of dirt, 4 drops of blue food coloring, and 5 splashes of water. Mix well.

Using a conversion factor, determine how many drops of blue food coloring should be used with 85 splashes of water.

(Start with the given 85 splashes of water. You want to "convert" that into drops of food coloring.)

6. 18 karat is probably the most common alloy of gold. What percent gold is that?

.......COMPLETE SOLUTIONS.......

1. Since we can see the different components, this is a heterogeneous mixture.

There are mixtures which are right between heterogeneous and homogeneous. If you start to grind up the sand and iron filings, at some point before you get a brown-gray powder, it will be almost impossible to distinguish the sand from the filings.

There are many other pairs of adjectives that are like this. Light/heavy or cheap/expensive or hot/cold or living/dead. In each case, it's easy to distinguish the extremes. The sun is hot and liquid helium is cold. With living/dead, there is no question that you are alive and George Washington is dead. But lawmakers continue to debate where the line is to be drawn between the extremes.

2. The easiest way that I can think of is to use a magnet. It will pull the iron filings out of the sand.

3. (1) Sticking a straw in and drinking the water is very similar to decanting.

(2) Heat the beaker up with a Bunsen burner. The water will boil off leaving the oil.

(3) Ignite the oil. Oil is an organic compound (contains carbon) and it will burn. It may create a big smoky mess. You might have to wash the walls and ceiling after you are done.

4. Use soap.

5. $\frac{\text{85 spl}\cancel{\text{ashes of water}}}{1} \times \frac{\text{3 drops of food coloring}}{\text{5 spl}\cancel{\text{ashes of water}}} =$

51 drops of food coloring

6. 18K = 18/24 = 3/4 = 75%

If Joe mixes up the dirt, blue food coloring, and water with his hands, his sky-blue mud pies will be perfect.

And his hands will be . . . blue.

Soap is really good at creating a friendship between oil and water. The molecules of soap can hold hands with both the oil and the H_2O molecules. Soap should get the Nobel prize for peace.

But it doesn't work so well on blue food coloring. ☹

Chapter Thirteen
Compounds

Darlene took the fishing magazine out of Joe's hands. She showed him the colors of the gold alloys—red gold, rose gold, white gold, gray-white gold, soft green gold, and purple gold—and asked him which one would look best on his blue hands.

"But didn't you hear the man?" Joe said. "He said those were just mixtures. I wouldn't want to wear any mixed-up bunch of metals."

Darlene didn't know how to answer that objection. All of the wedding rings were alloys. She only had three choices: mixtures (alloys), elements, or compounds.

She flipped to the periodic table of elements in her chem book. It listed 118 elements. Twenty-eight of them did not occur in nature but were only created in labs. (The math: 118 – 90 naturally occurring = 28)

She crossed off the gases, such as hydrogen, helium, nitrogen, oxygen, chlorine, and fluorine. Those wouldn't work for making wedding rings.

She crossed off the two elements that are liquid at room temperature: mercury (Hg) and bromine (Br).

Gallium (Ga) is a pretty gray metal, but it melts at 303.3K which is 30° C. (The math: 303.3 – 273 ≈ 30) The stuff will melt in your hand.* A wedding ring made out of Gallium would last as long as some marriages nowadays.

She gave up on using elements for making wedding rings. No mixtures, no elements—all that was left was compounds. She listened very carefully as Fred starting talking about compounds.

"Compounds consist of two or more different elements. Don't confuse them with the seven elements that exist as diatomic (two atoms) molecules: H_2, N_2, O_2, F_2 (fluorine), Cl_2, Br_2, and I_2 (iodine).

"The weird thing about compounds is that they are often totally different than the elements that they came from. The best known example

---

* Butter melts at 32–35° C (depending, I guess, on the cow) and body temperature is 37° C.

is water, H_2O. Hydrogen and oxygen are both gases. When you combine them together you really wouldn't expect to get . . . water. Fred wrote on the board the chemical equation:

$$2H_2 + O_2 \rightarrow 2H_2O$$

"Later we are going to show you how to balance this equation.*"

Darlene raised her hand. "Excuse me. [Joe never said those words.] In our math class you told us that it was important to recognize when we were working with an equation. Equations and expressions were treated differently." She pointed to the famous box in her Beginning Algebra book:

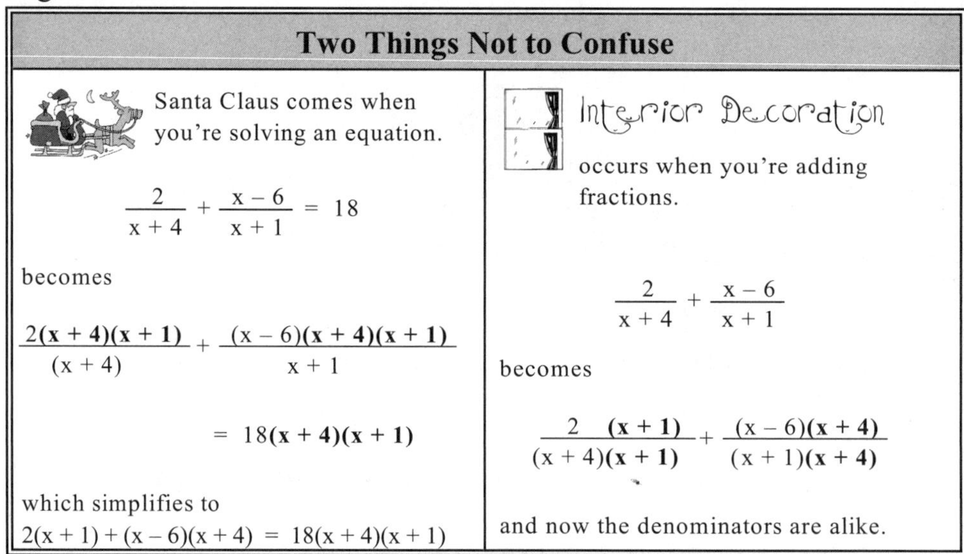

Two Things Not to Confuse

Santa Claus comes when you're solving an equation.

$$\frac{2}{x+4} + \frac{x-6}{x+1} = 18$$

becomes

$$\frac{2\mathbf{(x+4)(x+1)}}{(x+4)} + \frac{(x-6)\mathbf{(x+4)(x+1)}}{x+1} = 18\mathbf{(x+4)(x+1)}$$

which simplifies to

$2(x+1) + (x-6)(x+4) = 18(x+4)(x+1)$

Interior Decoration occurs when you're adding fractions.

$$\frac{2}{x+4} + \frac{x-6}{x+1}$$

becomes

$$\frac{2\ \mathbf{(x+1)}}{(x+4)\mathbf{(x+1)}} + \frac{(x-6)\mathbf{(x+4)}}{(x+1)\mathbf{(x+4)}}$$

and now the denominators are alike.

"You said that the way to recognize equations is that they have equal signs."

Fred nodded.

Darlene said, "But $2H_2 + O_2 \rightarrow 2H_2O$ doesn't have any equal signs, and yet you called it an equation."

"I called it a *chemical* equation. In math, equations have equal signs and you solve them. In chemistry, equations have → and you balance them.

---

* If you just wrote $H_2 + O_2 \rightarrow H_2O$, this equation would not be balanced. There are two atoms of oxygen on the left side and only one on the right.

"In math, the two sides of the equation are equal. In chemistry, the left side *becomes* the right side in a chemical reaction."

Darlene had hoped that she had caught Fred in an error. Fred gives 50 points to any student who catches him in an error at the board.*

Fred explained, "So $2H_2 + O_2 \rightarrow 2H_2O$ means that two molecules of hydrogen will combine with one molecule of oxygen and create two molecules of water.

"Oxygen likes to combine with lots of things. We call it burning. Heat up a piece of wood and the gases that come off the wood combine with the oxygen in the air."

Joe stood up and then sat down. He was excited. It suddenly made sense to him—why he saw flames when wood burned. It was gases that were burning. He turned to Darlene and said, "My mother never told me that. She just said that it was 'fire decoration'."

Of course, Joe had also never learned to use an "inside voice." He shouted in his "playground voice" and everyone in the room heard him. For weeks afterward, people teased Joe. When they saw clouds, they told him it was "sky decoration." Clothes were called "people decoration." And food was called "plate decoration."

Fred said, "Not all combining with oxygen results in flames. When oxygen combines with iron (Fe), it's called rusting. This happens especially quickly in the presence of water. He wrote on the board:

$$4Fe + 3O_2 \rightarrow 2Fe_2O_3$$

Darlene recognized why wedding rings aren't made from pure iron.

---

* It should be noted that in Fred's class, each quiz is worth 200,000 points and each test is worth 1,000,000 points. Fifty points sounds like a lot until you compare it to quizzes and tests.

When I was a college student, I took an upper-division (junior/senior) class in projective geometry. Our three-hour final exam was worth 10 points, and that was one-third of our final grade. That made the teacher's record keeping very simple.

Carpenters in the class knew that you don't use regular iron nails in outdoor construction where they can get wet. They use **galvanized** nails—nails coated with zinc. It's a dance. Lady oxygen would be delighted to dance with iron

or with zinc.

But zinc is much more willing to dance. In fact, if oxygen is already dancing with iron, zinc will cut in and take oxygen away from iron. Iron will remain unoxidized (unrusted).

In later courses in chemistry, numbers are assigned to iron's and zinc's dancing abilities—the **standard potential** for each of them. The standard potential for zillions of different dancing partners are often listed in big tables.

At one end of the table (the oxidizing side where oxygen is) are F_2 (fluorine) and O_3 (ozone).

At the other end are potassium (K) and lithium (Li).

When fluorine and lithium are dancing, nobody cuts in on them.

Darlene decided to buy a couple of T-shirts this afternoon with custom lettering on them. She was tired of seeing Joe's T-shirt that he wore all the time. It was a shirt his mother gave him years ago.

If Joe asked what "Lithium" meant, she would be happy to explain. Joe never asked.

He thought it was the name of a car.

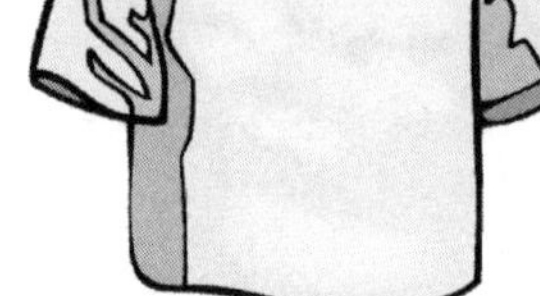

Your Turn to Play

H_2 O_2

1. If given the chance, two molecules of hydrogen will combine with one molecule of oxygen to form water. $2H_2 + O_2 \rightarrow 2H_2O$

The reaction goes from left to right.

We can force it to go the other way using electricity. A beaker full of water + two test tubes full of water + wires to a battery and we have **electrolysis**. (For English majors: *electro* = electricity + *lysis* = breakdown or decompose.)

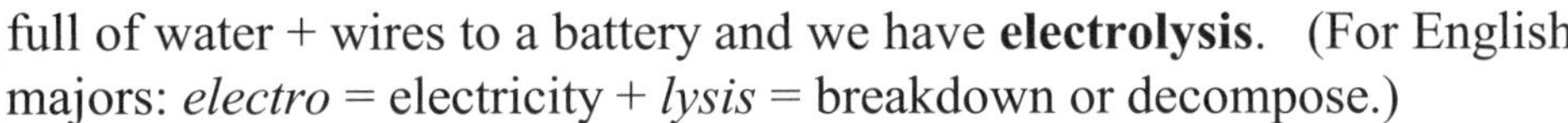

$$2H_2O \text{ (when shocked) } \rightarrow 2H_2 + O_2$$

Your question: If this experiment generates 44 mL of O_2, how much H_2 will be generated?

Small notes on this experiment. ♪#1: We use a battery, not current from a wall plug. A battery gives direct current. Electricity from an outlet is alternating current; it switches from + to – and from – to +. This would give H_2 and O_2 in both tubes. ♪#2: Using a bigger battery gives better results. ♪#3: Adding an inert electrolyte like $NaSO_4$ (sodium sulfate) into the water makes things go better. The sodium sulfate encourages the reaction without entering into it. ♪#4: I leave the insulation on the wires except for the centimeter of bare wire inside each test tube. That ensures that all the bubbles will be trapped by the test tubes.

2. $2H_2O$ (when shocked) $\rightarrow 2H_2 + O_2$ is an **endothermic** reaction. (For English majors: *endo* = goes in + *thermic* = regarding heat or energy.)

For endothermic reactions, you have to add energy (such as heat or electricity) in order to make the reaction happen.

Reactions that happen "happily" and that give off energy are called **exothermic** reactions. English majors might have guessed that.

Classify each of the follow as endothermic or exothermic.

A) Burning wood in a fireplace

B) Iron rusting $4Fe + 3O_2 \rightarrow 2Fe_2O_3$

C) Getting (most) teenagers to clean up their room

D) The major reaction that is going on in the sun

E) [harder question] The apricot that is ripening on my apricot tree.

.......COMPLETE SOLUTIONS.......

1. $2H_2O$ (when shocked) → $2H_2 + O_2$ takes two molecules of H_2O and produces two molecules of H_2 and one molecules of O_2.

These are gases.

Both the hydrogen and oxygen are at the same temperature and pressure.

A mole of hydrogen takes up the same volume as a mole of oxygen.

So if we generate 44 mL of O_2, we'll generate 88 mL of H_2.

2. A) Burning wood in a fireplace gives off heat. Exothermic

B) Iron rusts without having to encourage it. Iron doesn't "unrust"on its own. Exothermic

C) Mothers have to (usually) apply a great deal of pressure in order for this reaction (the room getting cleaned up) to happen. Endothermic

D) You may have noticed: the sun gives off a bunch of heat. In fact, it's the sun that keeps our earth nice and toasty. Exothermic

E) Put a tree in the dark and it eventually dies. Out in the sunshine its leaves enjoy the sun's energy. Carbon dioxide (CO_2) from the air and water from the wet soil are driven by the sunlight to make (among other things) sugars, such as those found in my apricots. This process is called **photosynthesis**. Endothermic

When you eat one of my apricots, your body breaks down the sugars into water and carbon dioxide. That reaction creates heat so that your internal temperature stays at 37°C. Digesting is an exothermic reaction.

The carbon dioxide you exhale was produced from my apricots. Joe's CO_2 was from his jelly beans.

Chapter Fourteen
Carbon Gas

Betty had a question. In contrast to Joe's questions, which are never very clear even in Joe's mind, Betty expressed herself well. Her questions could be complicated, but they were clear.

Betty began, "I know how we found the atomic mass of any of the gases. It's simple. You weigh 22.4 liters of the gas at standard temperature and pressure, and you're done. That weight in grams is the atomic mass. But how can you find the atomic mass of solids like carbon?" She pointed to the periodic table on the side of the lab table.

1 H 1.00794							2 He 4.00260
3 Li 6.941	4 Be 9.01218	5 B 10.811	6 C 12.0107	7 N 14.0067	8 O 15.9994	9 F 18.9984	10 Ne 20.1797
11 Na	12 Mg	13 Al	14 Si	15 P	16 S	17 Cl	18 Ar

"According to my chem handbook tables, carbon turns into a gas at 3370ºC. If I took the lead in my pencil, which is carbon, and tried to heat it up that much, it'd ignite long before it turned into gaseous carbon."

Fred liked questions like this. The first thing he needed to address was the problem of heating carbon without having it catch fire.*

---

* When I, your author, was about eight years old, I wondered whether it would be possible to melt wood. When you heated a log, the first thing that would happen would be for the water in the log to boil off. Water turning into steam and popping the space it's confined in may be why campfires crackle and pop. That's about as far as my eight-year-old imagination could take me.

Popcorn is corn with a hard shell. When you heat it, the little bit of moisture inside turns to steam and pops the kernel.

"A fire is a stool that has three legs,*" Fred said. "A fire needs (1) fuel, (2) oxygen, and (3) heat to get the fire started. If you take away any of these legs, the stool falls down—the fire goes out.

"If we are going to heat carbon to 3370ºC, we have the fuel and we have the heat. To prevent the fire we need to take away the oxygen.

"One approach would be to heat the carbon in a vacuum, but I don't have the vacuum equipment on this lab table.

"A second approach would be to heat the carbon in the presence of a gas that carbon won't react with."

"Hydrogen!" Joe suggested in his playground voice. He liked hydrogen because it was, as he liked to say, "Number one." It's atomic number—the number of protons—was one.

Fred wrote on the board:

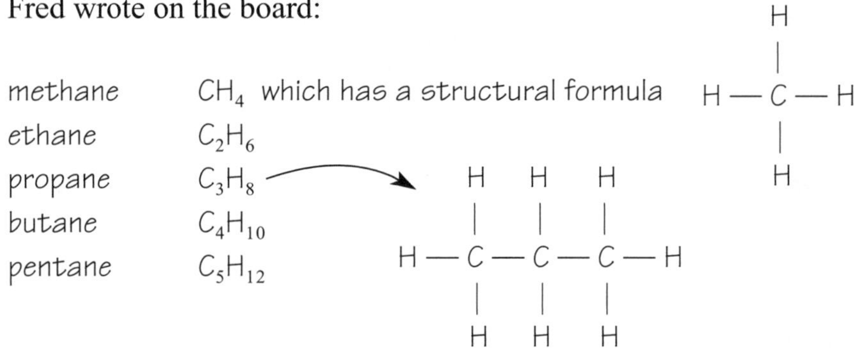

"The structural formulas show how the atoms are linked up in the molecule. We use a lot of structural formulas in organic chemistry."

---

* "A fire is a stool" is a metaphor. "A fire is like a stool" is a simile. "A fire requires three components" is boring and forgettable. There's no poetry in it.

Can you imagine writing a chemistry book that didn't have poetry, a story, and humor in it? That's impossible to imagine. This last sentence is ironic. Irony means writing the opposite of what we know to be true. It's much more fun to share in the joke of writing, "It's impossible to imagine a chemistry book without poetry, a story, and humor in it," than to write the boring truth: Virtually every chemistry book is dull.

Without directly saying that Joe was wrong, Fred indicated that heating carbon in the presence of hydrogen wouldn't give us carbon as a gas.

"There are six elements on the periodic table that really don't like to combine to make molecules. For two reasons they are called the **noble gases**. The first reason is that one of the meanings of *noble* is that it doesn't mix with others—standoffish. The second reason that those six elements are called noble gases is that they are . . . gases.

The six noble gases—helium, neon (Ne), argon (Ar), krypton (Kr), xenon (Xe), and radon (Rn)—are all found in the far right column of the periodic table. *Column* in math talk means vertical. *Row* means horizontal. A mathematician never says "vertical column" unless he or she is very sleepy or is talking to chemists.

2 He
10 Ne
18 Ar
36 Kr
54 Xe
86 Rn

Some chemists call this column **group VIII** or **group VIIIA**. Other chemists call it **group 18**. This is very much like how some people abbreviate a milliliter as mL, and some as cc, and some as cm^3.

Some chemists don't like to draw the lines of a structural formula. Instead, they will use double dots. They write methane as

```
    H                                H
    ‥                                |
H : C : H   instead of good old  H — C — H
    ‥                                |
    H                                H
```

They call those dotty things **Lewis structures**.

"So one approach would be for me to take some carbon, like pencil lead, and heat it up to 3370°C in the presence of helium. That would turn it into a gas. Measure the weight of 22.4 liters at standard temperature and pressure, and it would weigh 12.0107 grams—just like it says on the periodic table."

Fred looked out over the crowd. The television cameras were rolling. The reporters were taking notes. Darlene was looking at a bridal magazine, and Joe was eating a chocolate cupcake. "Can anyone see what the difficulty would be in carrying out that experiment?"

Joe raised his hand and said, "Ogla stumpful laytocklova sams." No one could understand him with his mouth full.

Darlene looked up and said, "The thermometer on your metal container only goes up to 200º C."

"That's easy," one of the reporters said. "Just wait till it all cools down to below 200º C."

"Duh," said Darlene.* "If you wait until it cools down to below 200º C, then it won't be a gas anymore."

Bob commented, "That Bunsen burner can't heat things up to 3370º C. According to the North American Combustion Handbook, volume 1, page 12, the flame temperature of propane/air is 1967º C."**

Fred wrote on the board:

$$C_3H_8 + 5O_2 \rightarrow 3CO_2 + 4H_2O \qquad \Delta H = -2020 \text{ kJ}$$

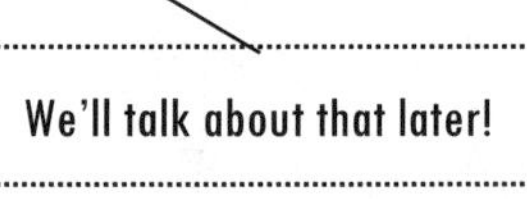

Bob continued, "So you want five moles of oxygen for every mole of C_3H_8 to have the best flame. A rich mixture—too much propane—and you get a yellow flame and the combustion is not complete. You get carbon soot in the air.

"If you have a lean mixture—too little propane—the temperature isn't as hot, and you get a bunch of hissing and popping."

"The best you can usually expect out of one of these burners is about 1760º C. This is chemistry and not math."

From the Handy Heat Hot sheet:

5700º C temperature at sun's surface
3370º C carbon turns to a gas
1760º C typical Bunsen burner temperature
1370º C steel **MELTS**

---

* Good teachers don't say "Duh" when students make suggestions.

** The North American Combustion Handbook is real. We try to keep the amount of fiction in the *Life of Fred* books to an absolute minimum. (The first sentence in this footnote is true.)

The whole metal container would melt if it contained carbon as a gas. Things were not looking very good for Fred if he had to find the atomic mass of carbon using only the stuff on Bob's lab table.

Fred said those famous words: "I will do it."

The class was utterly silent. Even Joe stopped eating.

These four words remain the most famous four words ever uttered in any chemistry textbook.

Your Turn to Play

1. Joe knew how to "help" Fred. He said, "If one burner can give you 1760° C, then you can easily get 3370°."

Fred asked, "How?"

Joe said, "Use two burners."

Your question: What did Darlene say at this point?

2. Write the structural formula and the Lewis structure for pentane.

(You are allowed to look back in the book, but please don't just turn to the next page and look at the answer. You will not learn as much if you do that.)

3. Here is the chemical equation that Fred wrote on the board for the combustion of propane: $C_3H_8 + 5O_2 \rightarrow 3CO_2 + 4H_2O$

If you want to know if the equation is balanced, you need to determine if:

A) the number of carbon atoms is the same on both sides of the equation,

B) the number of hydrogen atoms is the same, and

C) the number of oxygen atoms is the same.

Please make that determination.

4. Fred also wrote: $\Delta H = -2020$ kJ

Multiple choice question:

What does $\Delta H = -2020$ kJ mean?

A) Are you kidding?

B) I have no idea.

C) We've never covered that stuff.

D) You said you would talk about that later. Is "now" later?

.......COMPLETE SOLUTIONS.......

1. Duh.

2.

$$\begin{array}{ccccccccccc} & & H & & H & & H & & H & & H & \\ & & | & & | & & | & & | & & | & \\ H & — & C & — & C & — & C & — & C & — & C & — & H \\ & & | & & | & & | & & | & & | & \\ & & H & & H & & H & & H & & H \end{array}$$

$$\begin{array}{ccccccccccc} & & H & & H & & H & & H & & H & \\ & & \cdot\cdot & & \cdot\cdot & & \cdot\cdot & & \cdot\cdot & & \cdot\cdot & \\ H & : & C & : & C & : & C & : & C & : & C & : & H \\ & & \cdot\cdot & & \cdot\cdot & & \cdot\cdot & & \cdot\cdot & & \cdot\cdot & \\ & & H & & H & & H & & H & & H \end{array}$$

3. $C_3H_8 + 5O_2 \rightarrow 3CO_2 + 4H_2O$

A) There are three carbon atoms on each side.

B) There are eight hydrogen atoms on each side.

C) There are ten oxygen atoms on each side.

4. Any of the four choices is okay. A, B, C, or D.

$\Delta H = -2020$ kJ

I've got some explaining to do.

H stands for heat.

Δ stands for "change in."

ΔH means the change in heat. (You could have guessed that.)

k is the metric abbreviation for thousand. kg means 1000 grams.

J is a measure of heat. (pronounced JEWEL)

This is a Joule ⇨

THE LITTLE DETAILS

Joule's full name was James Prescott Joule, which his mom used when he was in trouble. On the playground (in the 1800s) everyone probably called him Jimmy Joule.

Heat is one form of energy. Officially, heat is the transfer of thermal energy.

A 100-watt light bulb uses 100 joules per second.

The negative sign in $\Delta H = -2020$ kJ means that the reaction is exothermic—throwing off heat. We would expect that when we burn propane.

If ΔH is positive, then the reaction is endothermic—it takes energy to run the reaction. You could have guessed that.

The genius readers at this point yell, "Wow!" Burn a mole of C_3H_8 (propane) means burning 44 grams (from the periodic table the math: $3\times12 + 8\times1 = 44$) gives off 2020 kJ which is 2,020,000 J which is about the heat of twenty thousand 100-watt light bulbs in one second. Hot!

Chapter Fifteen
Atomic Mass of Carbon

How could Fred do that? How could he find the atomic weight of carbon? There was no way he could turn carbon into a gas with the lab equipment he had. Even if he did, that hot gas would melt any container on that lab table.

Joe was willing to help. Simultaneously, he raised his hand and began talking, "I know how to find the atomic weight of carbon. It's right there on the periodic table."

Darlene hit him on the arm. "Don't you remember! Betty's question was how can you find the atomic mass. The 12.0107 on the table is the answer. She wants to know how to get that answer."

All eyes turned to Fred. Hollywood had sent a film crew to record this lecture. Next week, the movie would hit the theaters. "Life and Death in the Chem Lab" would star one of the young handsome actors. They would substitute his face over Fred's and stretch the film to make him look taller.

Add the usual Hollywood gratuitous sex and violence, delete some of the chemistry talk about moles (6.02×10^{23} of anything), add an Atomic Monster, and "Life and Death in the Chem Lab" will win all the Oscars.

It's the kind of movie that Darlene and Joe will love.

Meanwhile, Betty and Bob were eager to hear Fred.

"A second ago," Fred said, "I mentioned that the weird thing about compounds is that they are often totally different than the elements that they came from. It would be really hard to turn carbon into a gas, but carbon dioxide (CO_2) is a gas at room temperature. I could find the molecular weight of CO_2 by weighing 22.4 liters of it at standard temperature and pressure (STP). I already know how to find the weight of the elements that are gases at STP, such as H_2, He, N_2, and O_2.

"Once I find the weight of the CO_2, I just subtract the weight of the O_2, and I'm done."

Fred looked around on the chem table for a cylinder of CO_2. There wasn't any. He knew where to turn . . . to Joe. Darlene sat next to Joe. No one else sat near him. A perfect Circle of Mess surrounded him: candy wrappers, pink boxes of Waddles doughnuts, empty Sluice bottles, stacks of fishing magazines, cookie crumbs, and, of course, a popcorn machine.

"Joe," Fred asked, "do you have an extra bottle of Sluice? I could use the carbon dioxide bubbles for the experiment I'm about to do."*

Joe looked in his ice chest. "Sorry. I'm fresh out. I drunk the . . . "

Darlene, "*drank.*"

Joe, "I drank the last of the bottles a while ago."

Things looked pretty grim for Fred. There was no CO_2 available. He needed a gas that contained carbon. (Can you guess what he used?)

---

* A small note for those who take hot showers. Even though the temperature of the water is below boiling (100° C), the air is humid. It has water vapor in it. Warmer liquids have a higher **vapor pressure** than colder ones. Gasoline has a higher vapor pressure than water. It is more **volatile**. There are chemical handbooks that will give you the vapor pressures of lots of liquids at various temperatures.

If Fred were to use bubbles of carbon dioxide from Sluice, there would be molecules of water in that gas, and that would mess up his calculations.

One way to get rid of the water vapor is to use a **desiccant**. A desiccant is a solid that absorbs water. You sometimes find them in little packets or cylinders in vitamin containers with the words "Do Not Eat" printed on them.

Desiccant is a great spelling contest word. Many people miss the second *c* in the word.

It was right there on the table in front of him—the gas that is used in Bunsen burners—propane. C_3H_8

```
    H   H   H
    |   |   |
H — C — C — C — H
    |   |   |
    H   H   H
```

The only other atom in the molecule is hydrogen, and we knew its atomic weight since it was a gas at room temperature.

Fred detached the rubber tubing from the Bunsen burner and filled his metal container, blowing out the air so that pure propane was inside.

He did all the usual tricks that he had used to find the atomic weight of helium. The only new thing was that this was not the one liter glass container. He found the volume of this metal container by filling it with water and then pouring the water into a graduated cylinder, which is the chemist's form of a measuring cup found in kitchens.

Fred's final findings:

volume of the metal container = 1.7 L
temperature = 23° C
pressure = 1.3 atm
weight of the propane = 4.0126 g

Some people think that 90% of chem math is conversion factors. That's really silly.

It's closer to 95%.

Start with the weight of the propane. 4.0126 g. First, we adjust for the pressure. At 1.3 atm we pack in a lot more molecules than at 1 atm. So we are expecting less weight at a lower pressure.

$$\frac{4.0126 \text{ g at } 1.3 \text{ atm}}{1} \times \frac{\text{grams at } 1 \text{ atm}}{\text{grams at } 1.3 \text{ atm}} = 3.0866 \text{ g at } 1 \text{ atm}$$

Next we adjust for temperature. 23° C = 296.15 K
(The math: 23 + 273.15)
At the colder temperature of 0° C, there would be more molecules present (keeping the pressure constant). We can pack in more molecules when they are huddled together under the colder conditions.

$$\frac{3.0866 \text{ g at } 23^\circ \text{ C}}{1} \times \frac{296.15 \text{ K}}{273.15 \text{ K}} = 3.3465 \text{ g at STP}$$

The 3.3465 grams is the weight of the propane in 1.7 liters. We will expect more weight in 22.4 liters.

$$\frac{3.3465 \text{ g in } 1.7 \text{ liters}}{1} \times \frac{\text{in } 22.4 \text{ liters}}{\text{in } 1.7 \text{ liters}} = 44.095 \text{ g in } 22.4 \text{ L}$$

This is 44.095 grams of propane (C_3H_8) in 22.4 L at STP.
This is a mole of propane.

We know that a mole of hydrogen atoms weighs 1.00794 grams. We know this because we can determine the atomic weight of any element that is a gas at room temperature.

Eight moles of hydrogen weigh 8.06352 g (The math: 8 × 1.00794)

Then three moles of carbon weigh 36.0315 g
(The math: 44.095 – 8.06352 ≐ 36.0315)

Then one mole of carbon atoms weighs 12.0105 g.
(The math: 36.0315 ÷ 3) This is the atomic weight of carbon. (The definition of atomic weight is the number of grams that a mole of the stuff weighs.)

Since the pressure wasn't 1 atm but 1.7 atm,
since the temperature wasn't 0° C but 23° C,
since the volume was 22.4 L, but 1.7 L, we had to
make the appropriate adjustments.

Then we subtracted the weight of the 8 hydrogen atoms. C_3H_8
That left us with 3 carbon atoms. Then divide by 3.

It really is logical. Especially, after you've done a bunch of them.

Your Turn to Play

1. Sulfur (chemical symbol S) is often found on the chemist's shelf as a yellow, almost odorless powder. They say that is it tasteless, but that doesn't mean you should taste it. Sulfur's atomic number is 16, which means that each atom has 16 protons.

We want to find its atomic mass.

Hydrogen sulfide, H_2S, is a gas. It stinks. It really stinks. Like rotten eggs. Or flatulence (if you know what that means) that can knock over someone who's 20 feet away. The human nose can detect H_2S at concentrations of 0.02 ppm (parts per million). That's 2 atoms in 100,000,000 atoms of air.

(Historical note: One kid at my high school dropped a bit of FeS into some diluted sulfuric acid, H_2SO_4, and created a bit of H_2S gas. Everyone in the whole high school smelled it. By the way, at 100 ppm hydrogen sulfide creates sudden paralysis and death.)

All work with H_2S in a chem lab is done under a hood that prevents gases from entering the lab area.

Determine the atomic weight of sulfur. Here is the given information:

volume of the container = 1.7 L

temperature = 27° C

pressure = 0.6 atm

weight of the H_2S = 1.4123 g (Your answer should be about 32.)

2. Hey. Once we've found the atomic weights of some of the solids, then finding the atomic weights of other solids becomes a little easier. We don't have to hook up the solids to gases—as we did with carbon and sulfur.

The very first experiment that I did in my first chem class involved taking about a gram of copper wire and dumping a bunch of powdered sulfur on top of it and heating it in a crucible over a Bunsen burner. Some of the sulfur combined with the copper to make CuS. (Cu is the symbol for copper.) The rest of the sulfur burned off yielding SO_2, a gas. In the 1960s, no one used a hood. Nowadays, it's mandatory. Copper is bright and bendable. CuS is black and breaks apart if you mess with it; compounds are different than the elements that make them up. In my experiment, 1.43 g of Cu made 2.1515 g CuS. What's the atomic weight of copper. (You found the atomic weight of sulfur in problem 1.)

.......COMPLETE SOLUTIONS.......

1. We start with 1.4123 grams of H_2S.

The conversions for volume, temperature, and pressure can be done in any order. From algebra, you know that multiplication is commutative: ab = ba.

Volume: We know that 22.4 L will contain more molecules that 1.7 L.

$$\frac{1.4123 \text{ g in } 1.7 \text{ L}}{1} \times \frac{22.4 \text{ L}}{1.7 \text{ L}} = 18.609 \text{ g in } 22.4 \text{ L}$$

Temperature: We know that colder temperatures will contain more molecules than warmer temperatures.

$$\frac{18.609 \text{ g}}{1} \times \frac{300.15 \text{ K}}{273.15 \text{ K}} = 20.448 \text{ g at } 0^\circ \text{ C}$$

Pressure: Increasing the pressure (from 0.6 atm to 1 atm) will pack in more molecules.

$$\frac{20.448 \text{ g}}{1} \times \frac{1 \text{ atm}}{0.6 \text{ atm}} = 34.08 \text{ grams of } H_2S \text{ at STP}$$

The molecular weight of H_2S is 34.08.
The molecular weight of H_2 is 2.01588 (using the periodic table on p. 67).
(The math: 2×1.00794)
The atomic weight of S is 32.064 (The math: $34.08 - 2.01588 \doteq 32.064$).

2. One atom of copper combines with one atom of sulfur (yielding CuS). One mole of copper (6×10^{23} atoms) combines with one mole of sulfur.

This is a bit tricky. What to do next? Let's just start computing stuff. The CuS weighs 2.1515 g. The Cu weighs 1.43 g. The sulfur in the CuS must weigh 0.7215 g. (The math: $2.1515 - 1.43$)

So an atom of copper weighs 1.982 times as much as an atom of sulfur. (The math: $1.43 \div 0.7215 \doteq 1.982$)

So the atomic weight of copper is 63.55. (The math: 1.982×32.064)

Chapter Sixteen
Almost Nothing There

Fred had found the atomic weight of carbon just using everyday chem lab equipment. Dividing by the number of atoms in a mole, he had found the weight of a single carbon atom. (The math: 12.0105 g ÷ Avogadro's number = 12.0105 ÷ 602,000,000,000,000,000,000,000)

"I did it." ← *Perhaps the three most famous words ever uttered in any chemistry textbook.* (This was special. I know of no other chemistry textbook in which the hero takes regular lab equipment and determines the weight of a carbon atom.)

Betty said, "Thank you for answering my question."

Joe said, "I did it." These were the *least* famous words ever uttered in a chemistry textbook. Joe was referring to his having eaten everything he had brought with him. He borrowed Darlene's phone and ordered a pizza to be delivered from Stanthony's PieOne. He figured that would hold him over until his late morning snack.

Some of the television cameras had been focused on Joe's eating rather than on the chemistry magic that Fred had been performing. The television producers knew that the American public buys more cookbooks and more diet books than chemistry books.

One reporter interviewed Joe. She asked, "How are you able to consume those massive amounts of food? We saw you eat enough to completely fill 20 normal people."

Darlene thought That's easy. Joe is not normal. But she didn't say anything.

Joe explained, "Didn't you hear the lecture? Fred said that atoms are as far apart from each other as two tennis balls in five mansions. There's got to be lots of extra space in food."

The only one who hadn't been listening to the lecture was Joe. Darlene corrected him, "Fred said that was true for gas molecules—not for liquid or solid things."

Simultaneously, Joe raised his hand and and asked, "What about liquids and solids?"

Fred hadn't been listening to Joe's interview and couldn't make sense of "What about liquids and solids?"

"I mean, like, is there air between liquid atoms?"

Fred had a lot of explaining to do. “First of all, you mean is there much *space* between those atoms. Air is molecules, which are generally bigger than atoms.

“Secondly, atoms are neither solid, liquid, nor gas. Atoms are atoms. Being a liquid is a property of large collections of atoms or molecules. The same is true for color. You don’t have red copper atoms or blue silver atoms. They have color when you have so many of them that you can see them.”

Joe had gotten a lecture instead of a short answer. He asked, “Well, is there?”

“With liquids, the atoms or molecules are much closer together than with gases. And with solids they are (usually) even much closer. That’s why solids are denser—why it hurts much more to be hit by a hammer than by a blast of air.”

Betty asked a little different question. “What about inside of an atom? You said that most atoms are between 1Å and 5Å in diameter.” (1Å equals by definition 10^{-10} of a meter—one ten-millionth of a meter.)

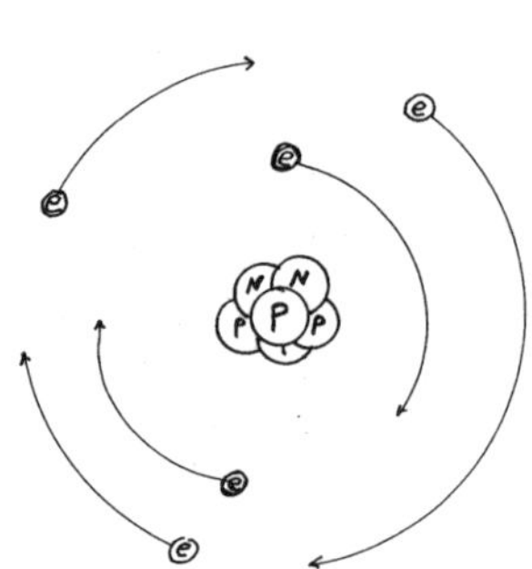

Betty pointed to the BASIC PICTURE of an atom that Fred had drawn on the board with the electrons circling the nucleus, which consisted of protons and neutrons. She asked, “Is there much air—I mean empty space—inside an atom?”

“I’m not that good at drawing,” Fred admitted. “If that famous artist, Kingie, were here, he would have made a much more accurate drawing.”

Joe’s pizza arrived. It was Stanthony’s personal size (18") combo pizza. The good smell (in contrast to H_2S) filled the room. Joe said, “What if a slice of pepperoni were the nucleus? Would this make a good picture of an atom?”

Fred asked, “Do you have anything smaller?”

“How about this little olive?” Joe rolled it around in the sauce with his finger.

“Perfect,” Fred said. “Now if that little olive were the nucleus then the whole atom with all the electrons flying around would be the size of—”

"The pizza!" Joe shouted.

"Not quite," Fred said. "If an atom's nucleus were the size of a small olive, then the atom would be the size of—"

"A mansion," Darlene said.

"Not quite. Your standard two-story mansion is only about 50 or 60 feet on each side if it were a square. Instead, think of the volume of that famous skiing hill.

Death Ski Hill

"Or of a big baseball stadium. That's the relationship between the olive-nucleus and the whole hill/stadium-atom.*"

Two tennis balls bouncing around in five mansions is crowded compared to an olive in a baseball stadium. Since 1836 electrons weigh as much as a proton, they contribute almost nothing to the mass of an atom. In a single five-word sentence:

---

* For people who like numbers more than ski slopes: The nucleus is about one-trillionth of the volume of the atom.

"How did they ever discover that the nucleus is so small?" Betty asked.

Fred dimmed the lights, pulled down the screen. Students loved it when Fred showed movies.

Invasion of the Beefy Atom

Fred Gauss Productions

A deep voice, two octaves below Fred's normal squeaky voice,* intoned: **A giant atom from outer space has come to conquer the world. It is nine feet tall and has six fingers on each hand.**

Only the Cowboy Kid could save the earth. He had to shoot Beefy Atom right through his little heart. After 20,000 shots, 19,999 of them going right through him harmlessly, he succeeded. THE END

"This is the way," Fred said, "that Ernest Rutherford in 1911 showed that the heart (the nucleus) of an atom is super small.

"Instead of a Beefy Atom monster, he used a thin sheet of gold. Gold is one of the most malleable elements. His gold foil was only a couple thousand atoms thick.

"Instead of Cowboy Kid's popgun, he shot alpha particles, which are about 7000 times heavier than electrons.

"Only about one out of every 20,000 alpha particles hit something heavy and was deflected by more than 90º. This showed Rutherford two things: (1) Except for the electrons buzzing around, atoms are mostly empty space and (2) there is an incredibly small and unbelievably dense *heart* in each atom."

---

* A musical note is a vibration at a certain frequency. If you want something an octave lower, you cut the frequency in half. The voice in the movie is really Fred's, but it is played at one-quarter of normal speed.

Your Turn to Play

1. Virtually all the weight of an atom is in the nucleus. (The number of electrons matches the number of protons, and protons are 1836 times heavier than electrons.)

Imagine an atom the size of Death Ski Hill. The nucleus would be about the size of a marble. How much would that marble weigh?

2. Red gold is an alloy of gold and copper. If I want 18 karat red gold and I start with 30 grams of gold, how much copper would be needed?

I have relatively few problems in comparison to many thousand-page chem textbooks that have 30 or 40 problems at the end of each chapter. Mine are few but choice. Doing 30 or 40 problems as fast as you can write is like working out with 3-ounce dumbbells.

This problem can be done in several different ways. It will probably take minutes, not seconds, to solve. Do you go to the gym and work out for three minutes and call that a real work out?

And looking at the problem and turning to the answer is even worse. That's like walking through a gym and just looking at the weights.

IN REAL LIFE, THE ONES WHO ARE PAID THE MOST
ARE THE ONES WHO SOLVE PROBLEMS.

3. Iron (Fe) and sulfur (S) combine to make pyrite (FeS_2) which is known as fool's gold. It looks like gold, but it isn't.

Remember Captain John Smith (and Pocahontas)? In the early 1600s, he was exploring the Chickahominy River (in the United States). He discovered what he thought was gold and shipped back *a whole shipload* of it to London. London wasn't very happy with John.

We want to find the atomic mass of iron. We know $Fe + 2S \rightarrow FeS_2$.

From the previous *Your Turn to Play* in problem #1, we found that the atomic mass of sulfur is 32.064.

In an experiment, 80.13 grams of iron combined with sulfur to create 172.14 grams of pyrite.

What is the atomic weight of iron?

Most readers take more than 5 minutes to solve this. There is no hurry. Take your time and enjoy it. We are striving for pleasant and challenging—not boring.

.......COMPLETE SOLUTIONS.......

1. Since virtually all the mass of an atom is in the nucleus, that marble would weigh as much as the whole hill.

Enlarge a nucleus to the size of a marble and put it on a six-inch thick steel plate and the force (tons!) of gravity would pull that marble right through it as the marble plunged toward the center of the earth.

Nuclei are **dense**.

2. There are many ways to find out how much copper would be needed. Here is one way. Eighteen karat gold means $\frac{18}{24}$ of the alloy is gold.

Reducing that fraction by dividing top and bottom by 6, we find that $\frac{3}{4}$ of the alloy is gold.

That means 3 parts of gold (Au) to 1 part of copper (Cu).

I like conversion factors.

$$\frac{30\text{ g }\cancel{\text{Au}}}{1} \times \frac{1\text{ }\cancel{\text{part}}\text{ Cu}}{3\text{ }\cancel{\text{parts}}\text{ }\cancel{\text{Au}}} = 10\text{ g Cu}$$

3. If you don't know where to start, you can just mess around with what was given and see what turns up.

We know that 80.13 g of Fe combined with sulfur to make 172.14 g of FeS_2. Then 92.01 g of sulfur must have been used. (The math: 172.14 – 80.13)

$Fe + 2S \rightarrow FeS$ tells me that one mole of iron reacted with two moles of sulfur.

A mole of sulfur weighs 32.064 grams. (That's what the atomic mass means.) Two moles of sulfur weigh 64.128 grams.

If you like algebra: $\frac{80.13\text{ g}}{92.01\text{ g}} = \frac{\frac{x\text{ g of Fe}}{\cancel{1\text{ mole of Fe}}}}{\frac{\cancel{2\text{ moles of S}}}{64.128\text{ g}}}$

Or without algebra: Two atoms of Fe would weigh 1.742 times as much as two atoms of sulfur. (The math: 2 × 80.13 ÷ 92.01)

So the atomic weight of iron is 55.86. (The math: 1.742 × 32.064)

Chapter Seventeen
Early Chemistry

Fred had reached a critical point in his chemistry lecture. Many generations ago people knew virtually nothing about chemistry (or math or astronomy or computers).*

Plato (about 400 B.C.) was the first that we know of to say that air, fire, earth, and water are the four elements from which everything is made. He obviously didn't know how to do electrolysis of water, which breaks it down into hydrogen and oxygen (as described on page 87).

Plato's student, Aristotle (about 350 B.C.), in his book *On Generation and Corruption*, advanced chemical knowledge by noting:

Fire is primarily hot and secondarily dry.
Air is primarily wet and secondarily hot.
Water is primarily cold and secondarily wet.
Earth is primarily dry and secondarily cold.

Whoopee! Now we're really getting somewhere. (Just kidding.)

Of course, both Plato and Aristotle were writing in Greek. They did that either (1) because they wanted to make it really hard for English-speaking readers to understand them, or (2) because they were Greek.

About 800 years later, Proclus (about A.D. 450) said that hot, cold, wet, and dry are stupid. Instead Proclus "really advanced" chemistry by telling his students that fire, air, water, and earth should be classified by sharp, dense, and immobile.

Fire is sharp. The rest aren't.
Earth is immobile. The rest aren't.
Water and earth are dense. The rest aren't.

The final exam in Proclus' chem course really wasn't that tough.

At this point in Fred's lecture, he had shown how to find the atomic weights of the elements. He had done this using really basic chemistry equipment. Fred's advantage over Plato, Aristotle, and Proclus was that he wasn't *telling* nature—"Fire is sharp"—as much as *asking* nature by performing experiments to find out what nature would tell him.

---

* This is not to say that they knew nothing. From the early records we can tell that they knew: (1) how to make babies; (2) how to kill each other; and (3) how to make wine.

Time Out!

There are fancier ways of finding the atomic weights of the elements. The important point is that Fred was able to do it using really simple lab equipment.

In the early 1900s, the **mass spectrometer** was invented. It is fancy and expensive.

Take an element and whack it with electrons (shot out of an electron gun. Yes, they call it an electron gun.)

Suppose the element you are whacking is sodium (Na). Sodium is known to have a loose electron that is easily detached from the atom.*

When the number of electrons don't match the number of protons, you become an **ion**. You are unbalanced.

Ions are unbalanced

The magnets in the spectrometer can now pull you around, and how fast you can make the turns tells how fat you are—your atomic mass.

It can even tell what percentage of each isotope is present in the sample. (Isotope = same number of protons but different number of neutrons)

Small note: If Plato put mass spectrometer on his Christmas list, he would be very silly. He would have to wait 400 years till the first Christmas. Then when he got his spectrometer, he couldn't plug it in.

---

*All the group 1 elements, the **alkali metals**, have this tendency to lose one of their electrons. Some people say that it's a bad habit they have.

The alkali metals are lithium (Li), sodium (Na), potasium (K), and some others you have never heard of.

When Fred teaches about the alkali metals, he likes to recite "Little Bo Peep has lost her sheep."

When he gets to the elements that tend to lose two electrons (the group 2 elements, the **alkaline earth metals**)—magnesium (Mg), calcium (Ca), and others—Fred will sing about the kitten who lost her mittens.

Fred had shown how to find atomic masses. The next step was to find atomic numbers (= the number of protons in an atom). He first listed the elements arranged in order of increasing weights:

H 1, He 4, Li 7, Be 9, B 11, C 12, N 14, O 16, F 19, Ne 20, Na 23, Mg 24, Al 27, Si 28, P 31, S 32, Cl 35, Ar 40,

K 39, Ca 40, Sc 45, Ti 48, V 51, Cr 52, Mn 55, Fe 56, Co 59, Ni 59, Cu 64, Zn 65, Ga 70, Ge 73, As 75,

Se 79, Br 80, Kr 84 . . . all the way up to Lv 293.

"The next step," Fred said, "is pretty easy. The atomic numbers are

H 1	He 4	Li 7	Be 9	B 11	C 12	N 14	O 16	F 19	Ne 20	Na 23	Mg 24	Al 27	Si 28 . . .
1	2	3	4	5	6	7	8	9	10	11	12	13	14 . . .

"The only tough thing was cobalt (Co) and nickel (Ni) which both have an atomic weight of about 59. When we list all the elements in the periodic table, the chemical properties of Ni will match up better with palladium (Pd) and platinum (Pt) than cobalt does. That will tell us in which order to put cobalt and nickel.

"The last step is to put this long line of elements into the periodic table. It is like fitting together the pieces of a jigsaw puzzle.

"When doing jigsaw puzzles, many people start with the pieces on the borders. The "border pieces" in the list of the chemical elements are the six noble gases. These are the only elements that are **inert**—they don't react; they don't form chemical compounds. They will make up the right edge of the periodic table. This column is group 18. →→→→

2 He
10 Ne
18 Ar
36 Kr
54 Xe
86 Rn

"Add one proton to each of the noble gases and you get

3 Li
11 Na
19 K
37 Rb
55 Cs
87 Fr

This is group 1. They are all soft, silvery metals. Each of them melts at low temperatures. They are the left edge of the periodic table. The first three are the most well known: lithium (Li), sodium (Na), and potassium (K). Group 1 = the alkali metals

"In contrast to the group 18 noble gases, which are basically dead, these alkali metals are **alive**."

Fred put on some goggles. He cut off a small chunk of potassium (K) and put it into a beaker of water.

Fireworks!

$2K + 2H_2O \rightarrow 2KOH + H_2$

ΔH = huge negative number

≺an exothermic reaction≻

Heat and light are given off.

The fish really wasn't in the beaker. That's just a drawing to show where the water is.

So much heat that it ignites the hydrogen which then burns in the air, and this gives off more heat and light.

after taking off the goggles

Fred was very happy that he had put on goggles before dropping the potassium into the water.

A negative value for ΔH means that energy is released during the reaction. If I melt an ice cube, I'm adding heat into the process. In this case ΔH would be positive.

This is all too easy. Chemists would not like it if everyone thought their subject was so simple that even Joe could understand it. Joe knows that if ΔH is negative you get heat/light/sparks/explosion. He knows that if ΔH is positive, you have to use a Bunsen burner to drive the reaction, such as the Cu + S experiment in problem 2 on page 99.

Let's make things harder without increasing your knowledge and understanding. Instead of ΔH—the change in energy or heat—we will call it **enthalpy change**. At a party, you can tell the one you are trying to impress, "My glass of Sluice has experienced a positive change in enthalpy" (because you were holding it in your hot little hand).

If they understand that, then go big time with **Hess's law**: *In a two-step process, you can add the* ΔH*s in each step to get the total* ΔH. The heat from adding K to water plus the heat from burning off the resulting H_2 gives the total heat given off. Any dummy knows that's true, but saying "Hess's law" makes you look really smart.

Your Turn to Play

1. This question is for those of you who might become writers some day. See if you can be as creative as the four readers who answered this question on the next page. Write your answer *before* you look at theirs.

Biological note: Your saliva (spit) is mostly water.

You like to chew gum? What do you think about the new K Sticks gum? Each stick is soft, silvery potassium.

2. Let's check to see if the chemical equation

$2K + 2H_2O \rightarrow 2KOH + H_2$ is balanced.

a) Are there the same number of K atoms on each side?

b) Are there the same number of O atoms on each side?

c) Are there the same number of H atoms on each side?

3. Sodium (Na) does a similar thing that potassium (K) does when it hits water. They are both alkali metals.

Have you ever in your life eaten something that contained a half gram of sodium atoms?

4. If you drop 5.00 grams of potassium (K) into water, how many grams of KOH (potassium hydroxide) are created?

Data that may be helpful:

atomic weight of K = 39.1 amu

atomic weight of O = 16.00 amu

atomic weight of H = 1.0079 amu

$2K + 2H_2O \rightarrow 2KOH + H_2$

5. If you drop 5.00 grams of potassium into water and the experiment is done in the presence of helium (He) instead of air (N_2 and O_2), then the hydrogen gas won't ignite. At STP what volume of H_2 will be made?

.......COMPLETE SOLUTIONS.......

1. Here are several readers' reactions:

#1: *No thank you. I'd rather not.*

#2: *I'd rather take up smoking . . . and getting drunk and then driving my motorcycle at night on icy roads.*

#3: *I hear that K Sticks gum are popular in Hell.*

#4: *I bet C. C. Coalback sells this kind of gum.*

2a) There are 2 atoms of K on each side.
2b) There are 2 atoms of O on each side.
2c) There are 4 atoms of H on each side.

3. On the first page of Chapter 13, Fred said, "The weird thing about compounds is that they are often totally different than the elements that they came from."

Take soft, silvery sodium metal and combine it with the poisonous pale yellow-green gas chlorine (Cl_2) and you get ordinary table salt.

$2Na + Cl_2 \rightarrow 2NaCl$ Imagine eating french fries without the salt. It would be like eating greasy potatoes.

4. $2K + 2H_2O \rightarrow 2KOH + H_2$ means that 2 atoms of K will make 2 molecules of KOH. It means that 2 moles of K will make 2 moles of KOH.

$$\frac{5.00 \text{ g of K}}{1} \times \frac{1 \text{ mole of K}}{39.1 \text{ g of K}} \times \frac{1 \text{ mole of KOH}}{1 \text{ mole of K}} \times \frac{56.1079 \text{ g of KOH}}{1 \text{ mole of KOH}}$$

≈ 7.1749 g of KOH $\doteq 7.17$ g of KOH

Or without conversion factors, you could note that 5 g of K is $\frac{5}{39.1}$ moles of K, which is 0.127877 moles. Then 0.12788 moles of KOH is (0.127877)(56.1079 g) $\doteq$ 7.17 g

5. $2K + 2H_2O \rightarrow 2KOH + H_2$ means that 2 moles of K will produce one mole of H_2. So 1 mole of K will produce 0.5 moles of H_2.

$$\frac{5.00 \text{ g of K}}{1} \times \frac{1 \text{ mole of K}}{39.1 \text{ g of K}} \times \frac{0.5 \text{ moles of } H_2}{1 \text{ mole of K}} \times \frac{22.4 \text{ L at STP}}{1 \text{ mole of } H_2}$$

≈ 1.4322 L $\doteq 1.43$ liters of H_2.

Note that for purposes of rounding, the 0.5 moles of H_2 and 1 mole of K are exact numbers with virtually an infinite number of significant digits. All the other numbers have 3 significant digits.

Chapter Eighteen
Creating the Periodic Table

Fred said, "Once you've got the left column with the alkali metals, and you've got the right column with the noble gases, you are almost done. All you have to do is string the rest of the elements between them. It's like hanging clothes on a clothesline."

Many of Fred's younger students knew what hard drives, USB ports, and address bars were, but few had ever hung clothes on a clothesline.

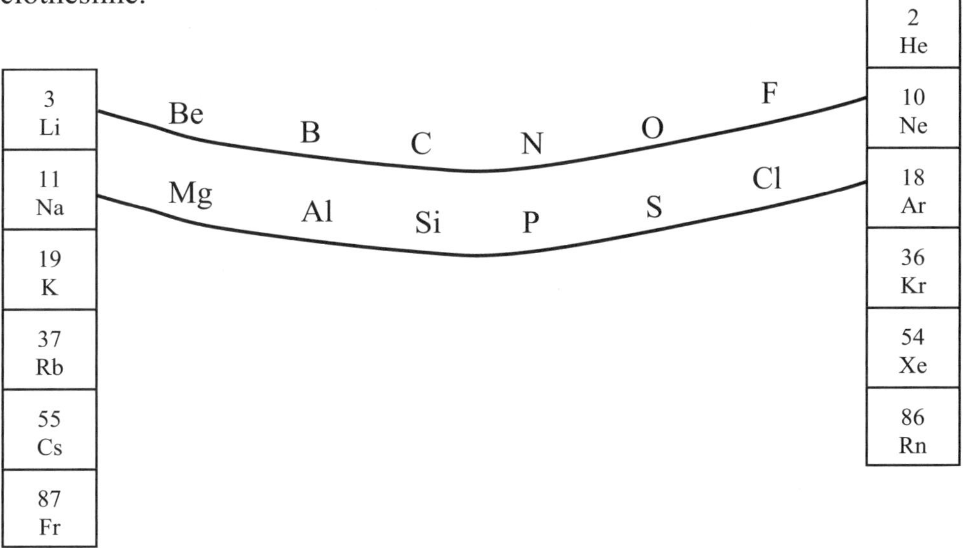

Fred pointed to the Periodic Table of the Elements poster on the side of the lab table and said, "Now you know how that was made."

1 H 1.00794							2 He 4.00260
3 Li 6.941	4 Be 9.01218	5 B 10.811	6 C 12.0107	7 N 14.0067	8 O 15.9994	9 F 18.9984	10 Ne 20.1797
11 Na	12 Mg	13 Al	14 Si	15 P	16 S	17 Cl	18 Ar

"If chemists were mathematicians, they would call the vertical stack of boxes columns. Instead chemists call them **groups**. Instead of calling the horizontal clotheslines rows, they call them **periods**.

"In the first period there are only two elements—hydrogen and helium. There is much debate as to whether H should be in group (column) 1 with the other alkali metals. Most people would not consider hydrogen as a soft, silvery metal, but they do stick it in group 1 because H has the same bad habit as sodium, potassium, rubidium, cesium, and francium. They all have the habit of losing one of their electrons and running around as ions: H^+, Na^+, K^+, Rb^+, Cs^+, and Fr^+."

When Fred strung his clothesline of Be B C N O F between Li (in group 1) and Ne (in group 18), it reminded Darlene of a necklace.

When he strung Mg Al Si P S Cl between Na and Ar (argon), those two periods (rows) of the periodic table became two strings of pearls—exactly what she might wear with her wedding dress.

She went to work stringing the next row of pearls. It would have to go between K (atomic number 19) and Kr (krypton, atomic number 36). Something went wrong. This fourth period (row) would have to have a lot more elements on the clothesline: Ca Sc Ti V Cr Mn Fe Co Ni Cu Zn Ga Ge As Se and Br. She put up her hand and said, "Excuse me. How?"

Fred asked, "How what?"

Darlene went to the board and drew in the next clothesline:

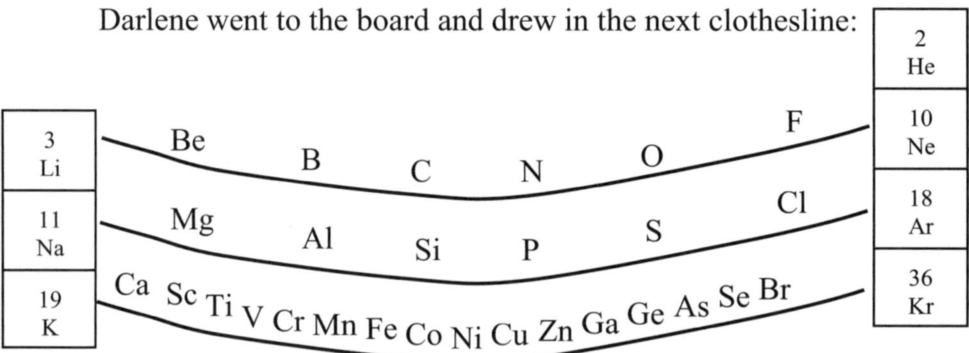

This periodic table became one big mess. It wasn't pretty anymore. The alkali metals (plus hydrogen) looked nice on the left side with their habit of losing an electron. The noble gases on the right all fit together with their habit of not being very sociable and not combining with anything to make compounds. But those pearls in the middle looked ugly.

"Instead of creating the periodic table from the top down," Fred said, "let's start with period 4 and work backward."

1	2	3	4	5	6	7	8	9	10	11	12	13	14	15	16	17	18
H																	He
Li																	Ne
Na																	Ar
K	Ca	Sc	Ti	V	Cr	Mn	Fe	Co	Ni	Cu	Zn	Ga	Ge	As	Se	Br	Kr

All of a sudden, many of the students realized why the noble gases were group 18.

"Filling in group 17 is the easiest spot to begin," Fred said. "That group really hangs together nicely. They are all colorful and corrosive:

- fluorine (F) pale yellow gas
- chlorine (Cl) yellow-green gas
- bromine (Br) red-brown liquid
- iodine (I) purple-black solid
- astatine (At) extremely rare and we don't know much about it

"These are the **halogens**. All of these guys go around looking for one more electron to make their lives complete."

Darlene thought of herself as a halogen. One electron (Joe) and her life would be complete.

"Those halogens will grab an electron wherever they can find one and become the well-known ions: F^-, Cl^-, Br^-, and I^-. They are electrically unbalanced, but very happy."

Fred filled in group 17:

1	2	3	4	5	6	7	8	9	10	11	12	13	14	15	16	17	18
H																	He
Li																F	Ne
Na																Cl	Ar
K	Ca	Sc	Ti	V	Cr	Mn	Fe	Co	Ni	Cu	Zn	Ga	Ge	As	Se	Br	Kr
																I	

And while he was at it, he filled in period 5.

1	2	3	4	5	6	7	8	9	10	11	12	13	14	15	16	17	18
H																	He
Li																F	Ne
Na																Cl	Ar
K	Ca	Sc	Ti	V	Cr	Mn	Fe	Co	Ni	Cu	Zn	Ga	Ge	As	Se	Br	Kr
Rb	Sr	Y	Zr	Nb	Mo	Tc	Ru	Rh	Pd	Ag	Cd	In	Sn	Sb	Te	I	Xe

Joe looked up (from his dim sum plate that he had been working on). "Hey. This is like doing a crossword puzzle."

For once, Joe was exactly right. Chemistry (and math and life) can most happily be approached as adventures rather than as stuff to be trudged through. Many traditional chem books hit their readers—splat!—with the whole periodic table all at once. It is like trying to eat a whole cow in one bite.*

It's insane to memorize all the chemical symbols. There are only a couple dozen that you use all the time. There are about five elements with weird symbols that are useful: Na (sodium), K (potassium), Hg (mercury), Au (gold), and Ag (silver).**

Fred broke into singing about the kitten who lost her mittens as he filled in the group 2 elements—the alkaline earth metals. These were the elements that tend to lose two electrons.

1	2	3	4	5	6	7	8	9	10	11	12	13	14	15	16	17	18
H																	He
Li	Be															F	Ne
Na	Mg															Cl	Ar
K	Ca	Sc	Ti	V	Cr	Mn	Fe	Co	Ni	Cu	Zn	Ga	Ge	As	Se	Br	Kr
Rb	Sr	Y	Zr	Nb	Mo	Tc	Ru	Rh	Pd	Ag	Cd	In	Sn	Sb	Te	I	Xe
	Ba																
	Ra																

Their ions look like Be^{2+}, Mg^{2+}, Ca^{2+}, etc.

---

* In fact, eating a whole cow in one bite is not pleasurable for either you or the cow.

** I have a personal favorite: tin (Sn). That's because Sn comes from the Latin word for tin: *stannum*. My name could be Sn Schmidt.

The next logical elements to fill in are those that *need* two electrons in order to be happy. We know that hydrogen, lithium, sodium, and potassium—all in group 1—like to lose a single electron.

We know that H_2O is world famous. (In fact, H_2O covers most of the surface of the earth.) Each H shares one electron. You could guess (correctly) that oxygen loves to receive two electrons. The ionic forms are O^{2-}, S^{2-}, Se^{2-}, etc. These elements fill in group 16.

1	2	3	4	5	6	7	8	9	10	11	12	13	14	15	16	17	18
H																	He
Li	Be														O	F	Ne
Na	Mg														S	Cl	Ar
K	Ca	Sc	Ti	V	Cr	Mn	Fe	Co	Ni	Cu	Zn	Ga	Ge	As	Se	Br	Kr
Rb	Sr	Y	Zr	Nb	Mo	Tc	Ru	Rh	Pd	Ag	Cd	In	Sn	Sb	Te	I	Xe
Cs	Ba														Po	At	Rn
Fr	Ra																

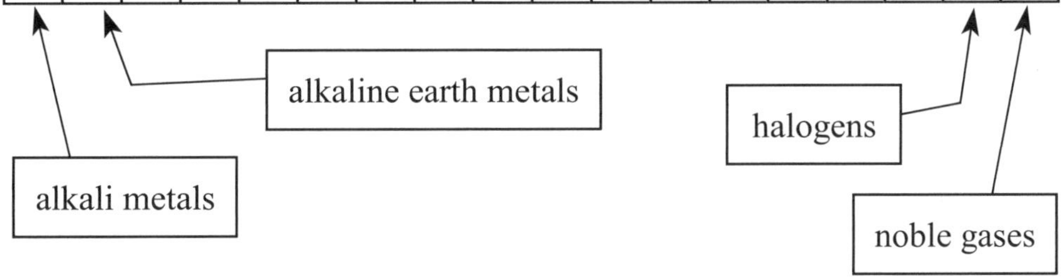

alkaline earth metals

halogens

alkali metals

noble gases

"Now things get a little more tricky," Fred said. "We have six elements in the original chart that we haven't stuck into the new chart that is 18 groups wide." He pointed to the original chart on the side of the lab table.

1 H 1.00794							2 He 4.00260
3 Li 6.941	4 Be 9.01218	5 B 10.811	6 C 12.0107	7 N 14.0067	8 O 15.9994	9 F 18.9984	10 Ne 20.1797
11 Na	12 Mg	13 Al	14 Si	15 P	16 S	17 Cl	18 Ar

"Those six have got to go somewhere in here."

1	2	3	4	5	6	7	8	9	10	11	12	13	14	15	16	17	18
H																	He
Li	Be														O	F	Ne
Na	Mg														S	Cl	Ar
K	Ca	Sc	Ti	V	Cr	Mn	Fe	Co	Ni	Cu	Zn	Ga	Ge	As	Se	Br	Kr
Rb	Sr	Y	Zr	Nb	Mo	Tc	Ru	Rh	Pd	Ag	Cd	In	Sn	Sb	Te	I	Xe
Cs	Ba														Po	At	Rn
Fr	Ra																

"The elements in groups 3–15 often can't make up their minds in regard to losing or gaining electrons.

"In contrast, Na^+ and K^+ have **oxidation states** of +1. F^- and Cl^- and Br^- have oxidation states of –1. They know exactly who they are and what they want. It made it easy to say that sodium and potassium are in group 1 and fluorine, chlorine, and bromine are in group 17.

"I'm trying to put nitrogen (N) on this chart. Nitrogen is wishy-washy. It can form the compound NO. Since oxygen is –2, the oxidation state of nitrogen in his compound must be +2.

"It can form NH_3. Since hydrogen is +1, the oxidation state of nitrogen in this compound must be –3. Oxidation states are sometimes called **oxidation numbers**."

Fred was getting frustrated with nitrogen. Aluminum was a lot nicer. $AlCl_3$ meant that aluminum had an oxidation state of +3 (since Cl^- has an oxidation state of –1). Al_2O_3 also meant that aluminum had an oxidation state of +3. Aluminum was acting like a metal (which is the left side of the periodic table. Aluminum looked like a metal. Aluminum conducted electricity (which is something that the non-metals on the right side of the table can't do very well). Aluminum *is* a metal.

Fred was all ready to stick Al in its logical spot.

1	2	3	4	5	6	7	8	9	10	11	12	13	14	15	16	17	18
H																	He
Li	Be														O	F	Ne
Na	Mg	☺													S	Cl	Ar
K	Ca	Sc	Ti	V	Cr	Mn	Fe	Co	Ni	Cu	Zn	Ga	Ge	As	Se	Br	Kr

Na^+(atomic number 11), Mg^{2+}(atomic number 12) followed by Al^{3+}(atomic number 13)

Bob Bunsen shook his head. ☹

Your Turn to Play

1. The noble gases—helium, neon, argon, krypton, etc.—are really happy where they are. They don't like to lose any of their electrons and they are not in the market to acquire any more electrons. What is their typical oxidation state?

2. In the first period (row), the **duet rule** applies: the happy state consists of two electrons. The Lewis structure for helium is He :

In the second period, the **octet rule** applies: eight is the magic number. Fluorine and neon are the in the second period.

The Lewis structure for neon is : Ne : (with a pair of dots above and below Ne)

Write the Lewis structure for fluorine (F).

3. Sodium (Na) lives on the left side of the periodic table with all the happy-go-lucky alkali metals (group 1). Unlike hard iron, you can cut lithium (Li), sodium, or potassium (K) with a dull knife.

Why would you prefer steel handcuffs (iron alloyed with a bit of carbon) to handcuffs made out of soft silvery sodium?

4. Fluorine, chlorine, and bromine are greedy. They will steal an electron from almost any victim they can find.

Lithium, sodium, and potassium who live on the other side of town (of the periodic table) are generous. They will happily give away an electron to just about anyone who asks.

If you are a little kid going around on Halloween asking for electrons and you go to the house of Mrs. Fluorine and say, "Trick or treat!" you had better beware. She will steal your candy! (your electron)

If you go to the door of Mrs. Sodium, you will have an electron in your candy bag even before you can finish saying, "Trick or . . . ."

If I tell you that someone is greedy (grabs electrons), where is she located on the periodic table and is she solid, liquid, or gaseous?

.......COMPLETE SOLUTIONS.......

1. The oxidation state of the noble gases is almost always zero.

Tiny exceptions: When you get far down into the periodic table, the atoms get really fat and with all those electrons zipping around, they have more trouble keeping track of everybody. Mothers with more than seven or eight kids sometimes confuse the names of their children.

Xenon is a big old noble gas in the fifth period (row). Young energetic fluorine in the second period is lean and hungry. It needs a single electron. Under special circumstances* fluorine will **attack**. The results are not very pretty. XeF_2, XeF_4, and even XeF_6 can result. This can almost bring tears to the eyes of chemists who are sensitive to this kind of injustice.

The oxidation number of Xe in XeF_4 is +4. Four precious electrons.

Fluorine is the only element that will attack xenon.

And it is a three-time offender: XeF_2, XeF_4, and XeF_6.

2. $\ddot{:\mathrm{F}}\,\underset{\cdot\cdot}{}\cdot$ As long as your answer had seven dots, it was okay.

3. In the previous chapter we saw what happened when an alkali metal (potassium) came in contact with water. There is usually a bit of moisture on your wrists—it's called sweat—especially when you are being arrested. Contact with sodium or potassium would result in great heat and maybe even some sparks. $2\,Na + 2\,H_2O \rightarrow 2\,NaOH + H_2$.

And the NaOH (sodium hydroxide) would also do nasty things to your skin. Have you ever had a clogged sink drain? Pour some sodium hydroxide solution down the drain and it will clear the drain by dissolving all the hair and other gunk that is clogging your drain. Do you remember potassium chewing gum? Sodium hydroxide shampoo is in the same category.

4. The elements that love to grab electrons are in groups 16 and 17 on the right side of the table. O, F, and Cl are gases. Br is liquid. S (sulfur) and Br are solids. (In contrast, *every* metal, except mercury, is a solid.)

---

* Professor Eldwood's *Great Chem Recipes*: To make a decent quantity of xenon tetrafluoride (serves six), mix one part of Xe with five parts of F_2 in a pressure cooker at 6 atm and 400° for a few hours.

This is an actual chemical procedure for making XeF_4. Just forget the part about "serves six" and forget about using a steel pressure cooker found in a kitchen. Fluorine (F_2) will **attack** almost anything it comes in contact with including the steel.

Chapter Nineteen
Finishing the Periodic Table

Fred trembled when Bob shook his head. In Fred's mind, aluminum just *had* to go in period (row) 3 and in group (column) 3. Bob went up to the board and filled in those six elements.

1	2	3	4	5	6	7	8	9	10	11	12	13	14	15	16	17	18
H																	He
Li	Be											B	C	N	O	F	Ne
Na	Mg											Al	Si	P	S	Cl	Ar
K	Ca	Sc	Ti	V	Cr	Mn	Fe	Co	Ni	Cu	Zn	Ga	Ge	As	Se	Br	Kr
Rb	Sr	Y	Zr	Nb	Mo	Tc	Ru	Rh	Pd	Ag	Cd	In	Sn	Sb	Te	I	Xe
Cs	Ba														Po	At	Rn
Fr	Ra																

It was Fred's turn to ask a question. "You have put aluminum, which everyone knows is a metal, way over on the right side of the table. How do chemists justify that?"

Bob smiled. "I did put aluminum with the metals. Look at Sn (tin), which sits in period 5, group 14. It's even farther right than aluminum. It turns out that most elements are generous with their electrons. Most elements are metals."

Bob draw a big black line through the chart.

1	2	3	4	5	6	7	8	9	10	11	12	13	14	15	16	17	18
H																	He
Li	Be											B	C	N	O	F	Ne
Na	Mg											Al	Si	P	S	Cl	Ar
K	Ca	Sc	Ti	V	Cr	Mn	Fe	Co	Ni	Cu	Zn	Ga	Ge	As	Se	Br	Kr
Rb	Sr	Y	Zr	Nb	Mo	Tc	Ru	Rh	Pd	Ag	Cd	In	Sn	Sb	Te	I	Xe
Cs	Ba														Po	At	Rn
Fr	Ra																

Bob said, "Everything to the left of this line is a metal."

Bob colored in seven boxes.

1	2	3	4	5	6	7	8	9	10	11	12	13	14	15	16	17	18
H																	He
Li	Be											B	C	N	O	F	Ne
Na	Mg											Al	Si	P	S	Cl	Ar
K	Ca	Sc	Ti	V	Cr	Mn	Fe	Co	Ni	Cu	Zn	Ga	Ge	As	Se	Br	Kr
Rb	Sr	Y	Zr	Nb	Mo	Tc	Ru	Rh	Pd	Ag	Cd	In	Sn	Sb	Te	I	Xe
Cs	Ba														Po	At	Rn
Fr	Ra																

"And these seven elements boron (B), silicon (Si), germanium (Ge), arsenic (As), antimony (Sb), telurium (Te), and astatine (At) are called **semimetals** or **metalloids**.

"The elements are not evenly distributed between metals, semimetals, and nonmetals. A great majority of the elements are metals."

Fred thanked Bob, and Bob headed back to his seat in the back of the auditorium classroom.

There are seven periods (rows) in the periodic table. These are the seven rows in the table that Fred had written on the board. All Fred had to do was insert the remaining two rows "of pearls" (as Darlene called them) and he would be done. **(Memory aid: pearls = periods)**

More grief awaited Fred as he stuck in the remaining elements.

	1	2	3	4	5	6	7	8	9	10	11	12	13	14	15	16	17	18
1	H																	He
2	Li	Be											B	C	N	O	F	Ne
3	Na	Mg											Al	Si	P	S	Cl	Ar
4	K	Ca	Sc	Ti	V	Cr	Mn	Fe	Co	Ni	Cu	Zn	Ga	Ge	As	Se	Br	Kr
5	Rb	Sr	Y	Zr	Nb	Mo	Tc	Ru	Rh	Pd	Ag	Cd	In	Sn	Sb	Te	I	Xe
6	Cs	Ba														Po	At	Rn
7	Fr	Ra																

In period 6, between Ba (barium, atomic number 56) and Po (polonium, atomic number 84) there were 27 elements!

They just didn't seem to fit.

	1	2	3	4	5	6	7	8	9	10	11	12	13	14	15	16	17	18
1	H																	He
2	Li	Be											B	C	N	O	F	Ne
3	Na	Mg											Al	Si	P	S	Cl	Ar
4	K	Ca	Sc	Ti	V	Cr	Mn	Fe	Co	Ni	Cu	Zn	Ga	Ge	As	Se	Br	Kr
5	Rb	Sr	Y	Zr	Nb	Mo	Tc	Ru	Rh	Pd	Ag	Cd	In	Sn	Sb	Te	I	Xe
6	Cs	Ba	La Ce Pr Nd Pm Sm Eu Gd Tb Dy Ho Er Tm Yb Lu Hf Ta W Re Os Ir Pt Au Hg Tl Pb Bi													Po	At	Rn
7	Fr	Ra																

When periods 4 and 5 were introduced, Fred had to widen the table. Now with periods 6 and 7, he had to do it for one last time. Thirty-two columns.

The Long Form of the Periodic Table

H																															He
Li	Be																									B	C	N	O	F	Ne
Na	Mg																									Al	Si	P	S	Cl	Ar
K	Ca															Sc	Ti	V	Cr	Mn	Fe	Co	Ni	Cu	Zn	Ga	Ge	As	Se	Br	Kr
Rb	Sr															Y	Zr	Nb	Mo	Tc	Ru	Th	Pd	Ag	Cd	In	Sn	Sb	Te	I	Xe
Cs	Ba	La	Ce	Pr	Nd	Pm	Sm	Eu	Gd	Tb	Dy	Ho	Er	Tm	Yb	Lu	Hf	Ta	W	Re	Os	Ir	Pt	Au	Hg	Tl	Pb	Bi	Po	At	Rn
Fr	Ra	Ac	Th	Pa	U	Np	Pu	Am	Cm	Bk	Cf	Es	Fm	Md	No	Lr	Rf	Db	Sg	Bh	Hs	Mt	Ds	Rg	Cn	?	Fl	?	Lv	?	?

"There it is . . . the periodic table of the elements. From these few elements will spring all the zillions of compounds that exist."

All these elements are a bit like people.

Some of them are well-known—almost everyone's heard of oxygen.

Some of them do important things—every living thing has carbon in it.

Some get around a lot—about 75% of the universe's matter is hydrogen.

Some are never found "outdoors" and only live in labs—every element from neptunium (Np) onward was created by humans.

Some are very rare—about 35 *atoms* of livermorium (Lv), the heaviest element in the table, have been created. They are not on display anywhere for several reasons: (1) You couldn't see a million atoms if they were all standing there together waving at you, (2) the longest-lived isotope of Lv has a half-life of 61 ms. (ms = milliseconds = thousandths of a second)

Some elements are so fat that their lifetimes are shortened—every element heavier than lead (Pb, sixth row, fifth from the right) is radioactive. They are so big that they fall apart just standing there waiting for the bus to come.

"But just like people—famous or known, fat or skinny—each of them is given one space on the periodic table. Undertakers call this the one-to-a-box rule. Preachers and the writer(s) of the Declaration of Independence tell us that all are created equal."

Meanwhile, Joe decided to actually take notes! When Darlene looked at his notes, she chided him, "Joe, you gotta learn to write smaller."

But even Darlene and Betty were having trouble getting period 6 to fit onto their papers.

Cs Ba La Ce Pr Nd Pm Sm Eu Gd Tb Dy Ho Er Tm Yb Lu Hf Ta W Re Os Ir Pt Au Hg Tl Pb Bi Po At Rn

Almost no one uses the Long Form of the Periodic Table. Instead, chemists chop out all those "extra elements" and stick them down below. The final result is this Short Form of the Periodic Table.

	1	2	3	4	5	6	7	8	9	10	11	12	13	14	15	16	17	18
1	1 H																	2 He
2	3 Li	4 Be											5 B	6 C	7 N	8 O	9 F	10 Ne
3	11 Na	12 Mg											13 Al	14 Si	15 P	16 S	17 Cl	18 Ar
4	19 K	20 Ca	21 Sc	22 Ti	23 V	24 Cr	25 Mn	26 Fe	27 Co	28 Ni	29 Cu	30 Zn	31 Ga	32 Ge	33 As	34 Se	35 Br	36 Kr
5	37 Rb	38 Sr	39 Y	40 Zr	41 Nb	42 Mo	43 Tc	44 Ru	45 Rh	46 Pd	47 Ag	48 Cd	49 In	50 Sn	51 Sb	52 Te	53 I	54 Xe
6	55 Cs	56 Ba	57 La	72 Hf	73 Ta	74 W	75 Re	76 Os	77 Ir	78 Pt	79 Au	80 Hg	81 Tl	82 Pb	83 Bi	84 Po	85 At	86 Rn
7	87 Fr	88 Ra	89 Ac	104 Rf	105 Db	106 Sg	107 Bh	108 Hs	109 Mt	110 Ds	111 Rg	112 Cn	113 ?	114 Fl	115 ?	116 Lv	117 ?	118 ?

leftovers from period 6	58 Ce	59 Pr	60 Nd	61 Pm	62 Sm	63 Eu	64 Gd	65 Tb	66 Dy	67 Ho	68 Er	69 Tm	70 Yb	71 Lu
leftovers from period 7	90 Th	91 Pa	92 U	93 Np	94 Pu	95 Am	96 Cm	97 Bk	98 Cf	99 Es	100 Fm	101 Md	102 No	103 Lr

Your Turn to Play

1. Those "leftovers" from period 6 are really unknowns.

Who has ever heard of cerium (Ce) or praseodymium (Pr) or neodymium (Nd)? ⇷ rhetorical question

The "leftovers" from period 7 has one element that you probably have heard of. Make a guess what U stands for.

2. This question may take a little detective work—mostly keen observation. The chemists couldn't call those two chunks that they cut out of the Long Form of the Periodic Table the "leftovers from period 6 and period 7." Instead they call those leftovers the **lanthanides** and the **actinides**.

Make a wild guess where those names came from.

.......COMPLETE SOLUTIONS.......

1. The atomic bomb made from uranium (U, atomic number 92) is associated with the date August 6, 1945. The plutonium (Pu, atomic number 94) bomb is associated with August 9, 1945. In a radio broadcast on August 15th, Emperor Hirohito announced the surrender of Japan.

2. One reader guessed that those leftovers were named after George Lanthanide and Mary Actinide. He was wrong. As far as I know, no one has ever had those names.

Instead, if we look at the spot in the Short Form of the Periodic Table where the leftovers were chopped out of . . .

6	55 Cs	56 Ba	57 La		72 Hf	73 Ta	74 W
7	87 Fr	88 Ra	89 Ac		104 Rf	105 Db	106 Sg

. . . the element just before the lanthanide series is La (lanthanum) and the element just before the actinide series is Ac (actinium).

Those chemists are sooo clever.

It might be helpful to place a bookmark on the previous page so that you can find the Short Form of the Periodic Table easily. Some people use a sticky note that hangs out the side edge of the book to mark this spot.

Chapter Twenty
Doing a Spearmint

Joe was getting a little impatient. All this stuff about the lanthanides and actinides wasn't the kind of chemistry that motivated/inspired/aroused/animated/galvanized him. With his unbandaged hand he stuffed several pink footballs of cotton candy into his mouth. Simultaneously, he raised his bandaged hand and asked, "When are you going to do a spearmint? That's the funnest part of chem."

Darlene hit Joe on the arm. "He just did an experiment. Where were you when he dropped that chunk of potassium into the water and made all those sparks and crackles?"

Joe shrugged his shoulders. He must have missed that.

Fred thought for a moment. He was trying to think of an experiment that the students could do, rather than one they could just watch him do. He knew that dropping potassium in water was definitely not a lab experiment for beginning chem students.

"Okay," Fred announced. "Everybody take your stuff. We're heading down the hallway to the chem lab. You're going to get the chance to do a real experiment."

As they walked, Fred thought of the traditional first experiment of heating up some copper wire with a bunch of sulfur on top. It's not very exciting, and all you get from the bright copper wire and the yellow powdery sulfur is some black piece of sludge that crumbles when you try to bend it. Fred didn't think Joe would enjoy that very much.

On the way down the hall, Joe filled his backpack with goodies from the 17 vending machines—9 on one side and 8 on the other—that lined the hallway of the Archimedes Building. (There are only nine vending machines on the third floor of the Math Building because that floor only has offices and not classrooms.)

As everyone assembled in the chem lab, Joe was working on a frozen treat. He knew enough *not* to put that in his backpack. And, besides, in his words he "needed the quick energy."

DANGER
DO NOT SMOKE, EAT OR DRINK IN THIS AREA

There were signs that Joe ignored.

DANGER
DO NOT SMOKE, EAT OR DRINK IN THIS AREA

Those signs were not posted so that the custodians would not have to clean up so much trash. In fact, they were busy right now cleaning up the "litter bomb" area that Joe had left behind in the classroom. They used shovels to scoop up all the pizza cartons, candy wrappers, and doughnut boxes that he left behind.

Those signs were in the chem lab so that students wouldn't get "killed to death."* There are chemicals in the lab that you wouldn't want to ingest. Even getting some of the things on your hands and subsequently on a cigarette would make smoking even less . . . well, you didn't want to see your grandkids graduate from high school, did you?

Everyone stood around waiting for Joe to finish his treat, so that they could get started on the experiment.

Fred began, "Your goal in this experiment is to determine the molecular formula for the compound made up from silver and chlorine. That compound is called silver chloride. The formula will look like $Ag_?Cl_?$. Your job is to find the value of those question marks.

"You may work alone or in pairs."

Darlene grabbed Joe's arm. Joe liked it when he was paired up with Darlene in the chem lab. That meant that he wouldn't have to pay as much attention to the chem stuff.

"At each lab station, you will find a small package of sodium chloride (NaCl), also known as table salt. You will find the usual assortment of beakers and funnels, a scale, and some filter paper. After Joe finishes eating, we may all begin."

After Joe finished his frozen treat, he reached into his backpack for some potato chips. Four students stopped him. Darlene told Joe, "If you'll stop eating now, we can go out to lunch after class is over. Okay?"

"Also, there is a bottle of silver nitrate at each lab station."

Fred pointed to the solubility chart on the wall.

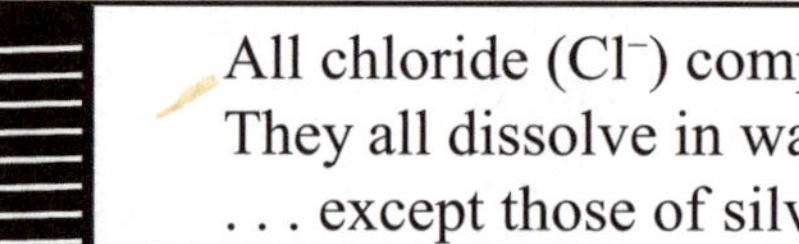

All chloride (Cl^-) compounds are soluble.
They all dissolve in water . . .
. . . except those of silver, mercury, and lead.

"Are there any questions?"

---

* Being "killed to death" was an expression my younger daughter used when she was small. When she was went to college, she was an English major.

One student: "What's silver nitrate?"

Fred pointed to a second solubility chart.

All nitrates (NO_3^-) are soluble.

A second student: "What's NO_3^- ?"

"It's an ion—a molecule of NO_3 that has gained an electron."

Everybody: "How do we do the experiment? You haven't told us."

Fred giggled. He knew that if he told them all the steps—first you do this; then you do this; then you do that—they could follow directions. It would just be like doing a recipe from a cookbook.

Fred is known world-wide as a Master Teacher. His students would not only have to *do* the experiment; they would have to *invent* it.

Bob Bunsen had never done this with his students. He had always told them exactly what to do. Bob's students would be good little workers at minimum wage. Fred's students would be owners or managers.

Chris and Pat filled a beaker with water, poured in the salt, stirred it, poured in the silver nitrate. A beautiful white cloud of silver chloride formed. (That's called a **precipitate**.) It settled to the bottom of the beaker. They decanted (see page 81) off most of the liquid. And they got stuck at this point.

Kim and Sidney did the same as Chris and Pat, except they put a piece of filter paper in a funnel and poured the everything into the funnel. They poured too quickly and some of the stuff flowed around the top edge of the filter.

Sam and Tracy did the same as Kim and Sidney, except that they poured more slowly and it didn't overflow around the top edge of the filter. But some of the precipitate stayed in the beaker. They didn't know how to push it out and into the filter.

Ashley and Kelly did the same as Sam and Tracy, except they just added some more water to the almost empty beaker, swirled it around and then could transfer it all to the filter paper. They dried the filter paper (to get rid of the water) and then weighed it. But they couldn't figure out what the net weight of the precipitate was because they couldn't get all of that silver chloride out of the filter paper.

Drew and Morgan did the same as Ashley and Kelly, except they weighed the filter paper before they poured the solution through it. Knowing the weight of the paper and the weight of the paper with the silver chloride, they could subtract to find weight of the precipitate. It was 0.344 grams. Drew and Morgan did a little victory dance.

They ran up to Fred and together said, "0.344 grams of silver chloride!"

Fred asked them, "Do you remember what you were looking for in this experiment?"

They looked at each other and shrugged their shoulders.

"You are trying to find the formula for silver chloride—find the value of the question marks in $Al_?Cl_?$."

Off in the corner of the lab Betty was working alone. Before she started weighing, pouring, filtering, and drying, she wrote out all the steps first. Then she wouldn't find herself stuck like Drew and Morgan were.

Betty first weighed the amount of salt (NaCl) that she poured into the beaker of water. 0.131 g of NaCl Then she stirred it. Poured in the silver nitrate and keep pouring until no more white clouds of silver chloride formed. At this point she knew that all the chloride from the salt had been precipitated out as silver chloride. Weighed the filter paper. Filtered the silver chloride. Dried the filter paper and precipitate and then weighed it. Subtracted the weight of the filter paper from the weight of the filter paper with the precipitate in it. The precipitate of silver chloride was 0.320 g.

Betty looked up the atomic weights in the periodic table: Cl is 35.45 amu; Na is 23.00 amu; Ag is 107.87 amu.

The 0.131 g of NaCl contains 0.07945 g of Cl.
from the experiment

$$\frac{0.131 \text{ g NaCl}}{1} \times \frac{35.45 \text{ amu Cl}}{58.45 \text{ amu NaCl}} = 0.07945 \text{ g Cl}$$

The 0.07945 g of Cl made 0.320 g of silver chloride.
from the experiment

The silver weighed 0.24055 g.

$$0.320 \text{ g} - 0.07945 \text{ g}$$

So 0.00224 moles of Cl

$$\frac{0.07945 \text{ g Cl}}{1} \times \frac{1 \text{ mole Cl}}{35.45 \text{ g Cl}} = 0.00224 \text{ moles Cl}$$

combined with 0.00223 moles of Ag.

$$\frac{0.24055 \text{ g Ag}}{1} \times \frac{1 \text{ mole Ag}}{107.87 \text{ g Ag}} = 0.00223 \text{ moles Ag}$$

An equal number of atoms of Cl and Ag. The formula is AgCl.

For most students, it will take about 20 minutes of rereading Betty's arithmetic before it starts to really make sense. Just reading her steps at the speed you read stories, will make the *Your Turn to Play* very difficult.

Your Turn to Play

1. Is lead nitrate $Pb(NO_3)_2$ soluble in water?

2. Dissolve 3 g of potassium iodide KI in water. Add a lot of lead nitrate. The precipitate (lead iodide) weighs 4.165 g.

Determine the value of the question marks in $Pb_?I_?$.

From the periodic table: I = 126.9 amu
Pb = 207.2 amu
K = 39.10 amu

Hints: We are going to follow the same steps that Betty did.

First, find the weight of the I in the KI.

Second, subtract that weight of the I from the lead iodide (4.165 grams) to find the weight of the Pb.

Third, convert the weights of the I and of the Pb into moles.

3. Everybody knows that mercury nitrate is soluble in water. [All nitrates are soluble.]

Take a large beaker with lots of mercury nitrate in water. Add 5 grams of NaI (which also dissolves nicely). A lovely orange precipitate of mercury iodide will form. It will weigh 7.574 grams.

Find the value of the question marks in $Hg_?I_?$.

From the periodic table: I = 126.9 amu
Hg = 200.6 amu
Na = 23.00 amu

.......COMPLETE SOLUTIONS.......

1. Look at the boxed solubility chart on page 129. All nitrates are soluble.
2. The 3 grams of KI contains 2.293 grams of I.

$$\frac{3 \text{ g KI}}{1} \times \frac{126.9 \text{ amu I}}{166 \text{ amu KI}} = 2.293 \text{ g of I}$$

The weight of the Pb is 1.872 g.

$$4.165 - 2.293 = 1.872$$

Moles of I = 0.01807 $2.293 \div 126.9 = 0.01807$ moles

Moles of Pb = 0.00903 $1.872 \div 207.2 = 0.00903$ moles

There are twice as many atoms of iodine as lead. PbI_2

3. The 5 grams of NaI contains 4.23 grams of I.

$$\frac{5 \text{ g NaI}}{1} \times \frac{126.9 \text{ amu I}}{149.9 \text{ amu NaI}} = 4.23 \text{ g I}$$

The weight of the Hg is 3.344 grams.

$$7.574 - 4.23 = 3.344$$

Moles of I = 0.03333 $4.23 \div 126.9 = 0.03333$

Moles of Hg = 0.01667 $3.344 \div 200.6 = 0.01667$

There are twice as many atoms of iodine as of mercury. HgI_2

In his lecture Fred first found the atomic weight of gases. Then atomic weights of the elements (such as carbon and sulfur) that formed gaseous compounds (CO_2 and HS). Then the atomic weights of each of the elements on the periodic table.

Then (in this chapter) he showed how to find the molecular formulas by measuring the weights of precipitates.

Knowing the atomic weights of elements and knowing the molecular formulas, gives us molecular weights.

Chapter Twenty-one
Tacos and Ducks

Betty had gotten to the "finish line." She had determined that the formula for silver chloride was AgCl. Bob Bunsen was worried that the other students hadn't learned anything because they hadn't gotten all the way through the experiment. When he did the teaching he told the students *exactly* what to do at every step. Most of Bob's students would be able to get to the final step and find the formula.

But they would have only learned how to follow directions. Fred's approach was to teach his students to think and to plan ahead.

small essay

The Story of Taco Quacko

A couple of years ago Fred had some free time when the university president had canceled classes. (In the Life of Fred math books this has been a recurring event.)

Fred had been reading *Life of Fred: Financial Choices*, and decided that he would start a business. One key element of a successful business is to offer something unique–something that the competition doesn't offer.

At that time Joe had been in Fred's arithmetic class. Every morning as Fred began that class, there would be a loud crunching sound as Joe took a big bite out of a hard-shell taco. Regardless of whatever else Joe ate during the lecture hour, he always started with a taco.

After class one day Fred had asked Joe, "Don't you ever get tired of eating those tacos every day?

Joe smiled. He knew that Fred didn't know much about eating. "There are a million* fast food taco joints out there. Some days I get a taco from one place and some days I get it from another place. That gives me variety."

To Fred, the taco that Joe had on Monday looked and smelled a lot like the one he ate on Tuesday, which looked a lot like the one on Wednesday and so on.

---

* This is hyperbole (high-PER-bow-lee)—an obvious exaggeration that is not meant to be taken literally.

The taco Joe had on Monday was a taco shell filled with ground beef, lettuce, cheese and a dash of sauce.

The taco Joe had on Tuesday was a taco shell filled with ground beef, lettuce, cheese and a dash of sauce.

The taco Joe had on Wednesday was a taco shell filled with ground beef, lettuce, cheese and a dash of sauce.

There was no real variety. Some were larger. Some were smaller. Some contained a little more meat.

Fred thought about all the taco poems he had read:

What Fred Remembered	The Original Poem
Oh my love is like a red, red, taco That's newly munched in June.	O my Luve's like a red, red rose, That's newly spring in June. Robert Burns, 1794
What's in a name that which we call a taco By any other name would taste as sweet?	What's in a name? that which we call a rose By any other name would smell as sweet. William Shakespeare
A taco is a taco is a taco is a taco.	Rose is a rose is a rose is a rose. Gertrude Stein, 1913

Fred's tacos would be different: Taco Quacko.

The building would be different.

The tacos would be in the shape of duckbills.

The workers would wear duckbill caps.

The advertising slogan would be **Dine at the Duck**.

end of small essay

The moral of the story is that students whose education consists of following directions and memorizing stuff, will most likely wear silly caps. When they see some trash on the floor of Taco Quacko, they will have to ask the manager, "What should I do?"

The manager will say, "Please clean it up."

They will ask, "How do you do that?"

The manager will demonstrate the use of a broom and dustpan.

In contrast, those students whose education emphasizes independence and learning to think will be those that Own a Piece of the Duck!℠ They will be owners or managers.

Small side note. Two-thirds of the university classes that I, your author, attended emphasized just memorizing stuff. Why do those teachers do that? Because having students memorize is *much easier for teachers* than doing the work of teaching them to reason.

Good little tape-recorder-brained students gobble up the facts thrown out by those teachers like ducks eating breadcrumbs.

In my art history class, for example, our whole job was to memorize the titles, painters, and dates for a bunch of paintings.

"Grace Allison McCurdy & Her Daughters, Mary Jane & Letitia Grace"
by Joshua Johnson
c. 1804

When I taught at the college level, I didn't want to do to my students what I hated when I was a student. Virtually all of the exams I gave were open book and open notes. I wanted my students to spend their time learning how to use the mathematics, not in memorizing that $\cos(x + y) = \cos x \cos y - \sin x \sin y$ or that the derivative of tan x is equal to $\sec^2 x$.

It's more work to teach that way, but I believe fewer of my students have to wear the hat.

"Is the spearmint over?" Joe asked Darlene as everyone headed back to Fred's auditorium classroom.

Bob had noticed that Chris & Pat, Kim & Sidney, Sam & Tracy, Ashley & Kelly, and Drew & Morgan had not been able to complete the experiment. What he failed to notice was that those ten students had learned a lot from their failure. Those ten will remember this experiment much longer than if they had just followed the cookbook steps that Bob might have laid out for them.

Back at the board Fred wrote the last words from Betty's experiment report: An equal number of atoms of Cl and Ag. The formula is AgCl.

"This is not quite right," Fred explained. "AgCl is the **empirical formula**. All we know from the experiment is that the number of atoms of silver equals the number of atoms of chlorine. The molecular (actual) formula might be Ag_2Cl_2 or $Ag_{57}Cl_{57}$.

"We encountered that with butane. (back on page 90)

The molecular formula is C_4H_{10}

The empirical formula is C_2H_5.

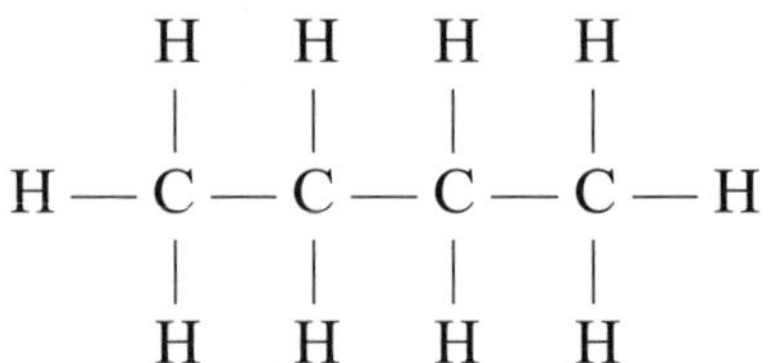

what butane really looks like

Kim thought that was a large chocolate bar.

Joe was starting to feel a bit weak. It had been minutes since he had eaten. He took out a candy bar and broke it into one-pound bites and ate the whole thing in two minutes and 42 seconds. His record is two minutes and 15 seconds, but he did that on a day he was hungrier.

Sugar ($C_6H_{12}O_6$) is oxidized in the body and creates carbon dioxide and water. The carbon dioxide is exhaled from the lungs. The balanced equation is

$$C_6H_{12}O_6 + 6\,O_2 \rightarrow 6\,CO_2 + 6\,H_2O$$

Fred answered a series of easy questions.

Q: One molecule of sugar will generate how many molecules of carbon dioxide.

A: Look at the coefficients of the balanced equation. One will produce six.

Q: One mole of $C_6H_{12}O_6$ will produce how many moles of CO_2?

A: A mole of *anything* is 6.02×10^{23}. Each molecule of sugar produces six molecules of carbon dioxide. One thousand molecules of sugar will produce six thousand molecules of carbon dioxide. One mole of sugar will produce six moles of carbon dioxide.

Memory Aids

When you are working with weights, convert things to moles.

When you are working with volumes, use 22.4 L of gas at STP equals one mole.

Your Turn to Play

1. Eight pounds of sugar (from that chocolate bar) will produce how many pounds of CO_2?

Atomic weights: H = 1.008
C = 12.01
O = 16.00

2. How many pounds of carbon is in that carbon dioxide?

3. How many moles of carbon is in that carbon dioxide?
Handy conversion: 1 pound ≈ 453.6 grams

4. What volume will the CO_2 occupy at standard temperature and pressure (STP)?

5. The empirical formula for butane is C_2H_5. It is a liquid at room temperature but is easily turned into a gas. How could you determine that the molecular formula is really C_4H_{10} and not C_2H_5?

6. Confirm that $C_6H_{12}O_6 + 6O_2 \rightarrow 6CO_2 + 6H_2O$ is balanced.

a) Are there the same number of C atoms on each side?
b) Are there the same number of H atoms on each side?
c) Are there the same number of O atoms on each side?

.......COMPLETE SOLUTIONS.......

1. When you are working with weights, first convert to moles.
The molecular weight of $C_6H_{12}O_6$ is 180.156 (The math: 6(12.01) + 12(1.008) + 6(16) = 180.156)
The molecular weight of CO_2 is 44.01 (The math: 12.01 + 2(16) = 44.01)
180.156 g of sugar will produce 6(44.01) g of carbon dioxide.

Using conversion factors . . .

$$\frac{8 \text{ pounds sugar}}{1} \times \frac{6(44.01) \text{ g carbon dioxide}}{180.156 \text{ g sugar}} = 11.73 \text{ pounds } CO_2$$

2. $\frac{11.73 \text{ lbs. of } CO_2}{1} \times \frac{12.01 \text{ amu C}}{44.01 \text{ amu } CO_2} \doteq 3.20 \text{ pounds C}$

3. $\frac{3.20 \text{ lbs. C}}{1} \times \frac{453.6 \text{ g}}{1 \text{ lb.}} \times \frac{1 \text{ mole C}}{12.01 \text{ g C}} \doteq 120.9 \text{ moles C}$

4. $\frac{120.9 \text{ moles C}}{1} \times \frac{22.4 \text{ L}}{1 \text{ mole}} = 2708.16 \text{ L} \doteq 2710. \text{ L}$

5. Heat it up and turn it into a gas. Measure the weight of the gas, its temperature and pressure. Using conversion factors find what the weight of 22.L of the gas would be at STP.

If it's 29.06 g (the molecular weight of C_2H_5) then C_2H_5 is the molecular formula.

If it's 58.12 g (the molecular weight of C_4H_{10}) then it's C_4H_{10}.

$$C_6H_{12}O_6 + 6O_2 \rightarrow 6CO_2 + 6H_2O$$

6a) There are 6 atoms of C on each side.
6b) There are 12 atoms of H on each side.
6c) There are 18 atoms of O on each side.

Chapter Twenty-two
Balancing Equations

Being able to determine how much carbon dioxide Joe would exhale after eating his family-sized chocolate bar* depends on having a *balanced* chemical equation: $C_6H_{12}O_6 + 6\,O_2 \rightarrow 6\,CO_2 + 6\,H_2O$

The **skeleton equation** is $C_6H_{12}O_6 + O_2 \rightarrow CO_2 + H_2O$ (where all the coefficients are 1). To balance an equation is to figure out what the coefficients must be so that there are the same number of each atom on each side.

In the skeleton equation $C_6H_{12}O_6 + O_2 \rightarrow CO_2 + H_2O$ there are six atoms of C on the left side and only one atom of C on the right side.

Fred started with a really simple example. On the board he wrote $H_2 + O_2 \rightarrow H_2O$ and asked if this skeleton equation was balanced.

Joe said, "Sure. There are two atoms of H on each side."

Darlene went to rescue Joe, "What he means is that even though there are an equal number of H on both sides, the equation isn't balanced because there are two atoms of O on the left side and only one atom of O on the right side."

Joe said, "Yeah. That's what I meant."

"How can we balance it?" Fred asked.

Joe tried again, "It's hard to say it. Can I just write it on the board?"

"Yes, you may," Fred answered, slightly correcting his English.

Joe altered Fred's equation $H_2 + O_2 \rightarrow H_2O_2$

Bob Bunsen had a difficult time watching Fred teach. When he was teaching students how to balance chemical equations, he just told them the steps. That was much easier than allowing the students to work it through and make mistakes. No one ever made mistakes in Bob's class because Bob never would let Joe talk. Bob's students were all good little soldiers who listened and obeyed. Soldiers with caps.

---

* Actually, few families could consume 15 pounds of chocolate.

Fred said, "$H_2 + O_2 \rightarrow H_2O_2$ is the correct equation . . . if you want to make hydrogen peroxide. Hydrogen peroxide (H_2O_2) is very different from water. You can buy a bottle of hydrogen peroxide in a drug store. It's used for sterilizing wounds and for bleaching hair. Almost anyone can* become a blond."

H_2O = okay to drink
H_2O_2 = don't drink it!

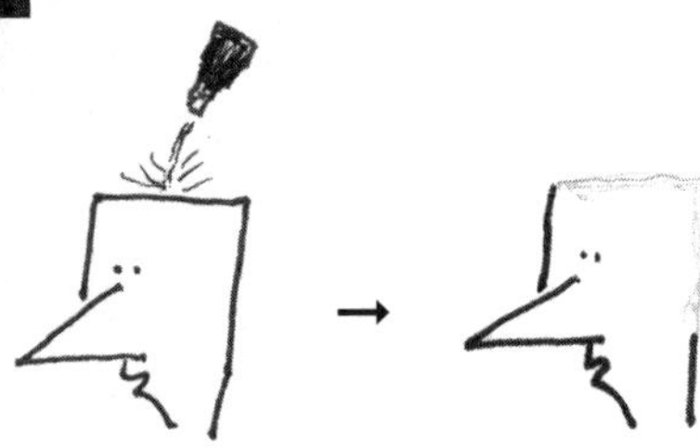

Fred once tried using H_2O_2.
But never again.

Fred erased Joe's subscript. $H_2 + O_2 \rightarrow H_2O$

Then he added some question marks. $?H_2 + ?O_2 \rightarrow ?H_2O$

"We are trying to find the coefficients. In the chem experiment you just did, you were trying to find subscripts. $Ag_?Cl_?$ That was for finding the molecular formula. Now we are trying to balance equations."

Joe took out a fresh sheet of paper and took notes.

big ?
little ?
I need a cookie.

Fred showed how to balance the equation.

We start with $H_2 + O_2 \rightarrow H_2O$
The H are balanced.
There are two O on the left side and only one on the right.
To balance the O $H_2 + O_2 \rightarrow \mathbf{2}H_2O$

* For some people bleaching would not make much sense.

Now the Hs are no longer balanced. $H_2 + O_2 \rightarrow \mathbf{2}H_2O$

That's easy to fix. $\mathbf{2}H_2 + O_2 \rightarrow \mathbf{2}H_2O$

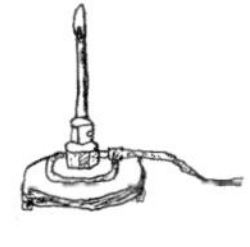

Fred fired up a Bunsen burner. "We are burning propane. Burning means combining with oxygen." He wrote the skeleton equation on the board.

$$C_3H_8 + O_2 \rightarrow CO_2 + H_2O$$

Chris & Pat, Kim & Sidney, Sam & Tracy, Ashley & Kelly, and Drew & Morgan were starting to twitch and wiggle.

Chris: "What rules do we follow to balance equations?"
Pat: "Tell us the steps."
Kim: "What do we do first?"
Sidney: "And then what?"
Sam: "What is the procedure?"
Tracy: "Tell me what to do and I will follow your instructions."
Ashley: "Lay it out so I can just do it."
Kelly: "I need to know in 1, 2, 3 order."
Drew: "Tell me exactly what to do."
Morgan: "Have you seen my new cap?"

Betty said, "If you just balance of a couple of equations, I bet we can figure out how to do it."

> "In this world, if a man sits down to think, he is immediately asked if he has the headache."
> —Ralph Waldo Emerson (1803–1882)

Fred liked Betty's approach. He wrote:

We start with $C_3H_8 + O_2 \rightarrow CO_2 + H_2O$

I like to look at the most complex molecule first. In this case, C_3H_8. It has 3 atoms of carbon. The right side has only one.

$$C_3H_8 + O_2 \rightarrow \mathbf{3}CO_2 + H_2O$$

There are 8 atoms of hydrogen on the left. Only 2 on the right.

$$C_3H_8 + O_2 \rightarrow \mathbf{3}CO_2 + \mathbf{4}H_2O$$

The C and the H are balanced. All that is left is the oxygen. There are 2 atoms of oxygen on the left and 10 on the right.

That's easy to fix. $C_3H_8 + \mathbf{5}O_2 \rightarrow \mathbf{3}CO_2 + \mathbf{4}H_2O$

Are we done? 3 C on both sides. 8 H on both sides. 10 O on both sides. Yes.

Fred warned the Trembling Ten (Chris & Pat, Kim & Sidney, Sam & Tracy, Ashley & Kelly, and Drew & Morgan) that sometimes you have to go back and forth, playing with the coefficients several times to make everything balance. "That's okay. It's to be expected. This is not as hard as trying to figure out the numbers for a combination lock."

Fred presented a third example.

Start with the skeleton equation (B is boron and P is phosphorus.)	$BCl_3 + P_4 + H_2 \rightarrow BP + HCl$
There are 4 phosphorus atoms on the left so I'll make 4 on the right	$BCl_3 + P_4 + H_2 \rightarrow \mathbf{4}BP + HCl$
2 H on the left and only 1 H on the right	$BCl_3 + P_4 + H_2 \rightarrow \mathbf{4}BP + \mathbf{2}HCl$
4B on the right and only 1 B on the left	$\mathbf{4}BCl_3 + P_4 + H_2 \rightarrow \mathbf{4}BP + \mathbf{2}HCl$
12 Cl on the left and only 2 Cl on the right	$\mathbf{4}BCl_3 + P_4 + H_2 \rightarrow \mathbf{4}BP + \mathbf{12}HCl$
That messed up the H which were in balance before	$\mathbf{4}BCl_3 + P_4 + \mathbf{6}H_2 \rightarrow \mathbf{4}BP + \mathbf{12}HCl$

Balanced! Each side has 12 Cl, 4 B, 4 P, and 12 H.

"The path from the skeleton equation to the balanced equation is not unique. For example, you might have started by balancing the Ps first. Then the first lines of your balancing act would have looked like this."

Start with the skeleton equation	$BCl_3 + P_4 + H_2 \rightarrow BP + HCl$
Balance the P	$BCl_3 + P_4 + H_2 \rightarrow \mathbf{4}BP + HCl$
Then the B	$\mathbf{4}BCl_3 + P_4 + H_2 \rightarrow \mathbf{4}BP + HCl$

and so on.

"In any event, you'll all finish in the same place. Sometimes you might overdo it a bit and end up with something like $20BCl_3 + 5P_4 + 30H_2 \rightarrow 20BP + 60HCl$, but dividing through by a number that divides into all the coefficients (in this case, 5) gives $4BCl_3 + P_4 + 6H_2 \rightarrow 4BP + 12HCl$, which is the final answer.

Your Turn to Play

Some chemistry students find that balancing chem equations is the most satisfying part of the course.

Balance these skclcton equations.

1. $P_4 + O_2 \rightarrow P_4O_{10}$

2. $IBr + NH_3 \rightarrow NI_3 + NH_4Br$ (Recall Fred's words from two pages ago: "I like to look at the most complex molecule first." In this case it is the NH_4Br.)

3. $KrF_2 + H_2O \rightarrow Kr + O_2 + HF$ (This is one of the very rare cases in which a noble gas—in this case, krypton, Kr—forms a compound. Even in the presence of water it falls apart.)

4. $Li_3N \rightarrow Li + N_2$

5. $BCl_3 + H_2O \rightarrow H_3BO_3 + HCl$

6. $Al + HCl \rightarrow AlCl_3 + H_2$ (Al is aluminum.)

Some chemistry students think that a mole is a small animal with tiny eyes that loves to dig tunnels in your lawn.

One mole of moles equals 6.02×10^{23} moles.

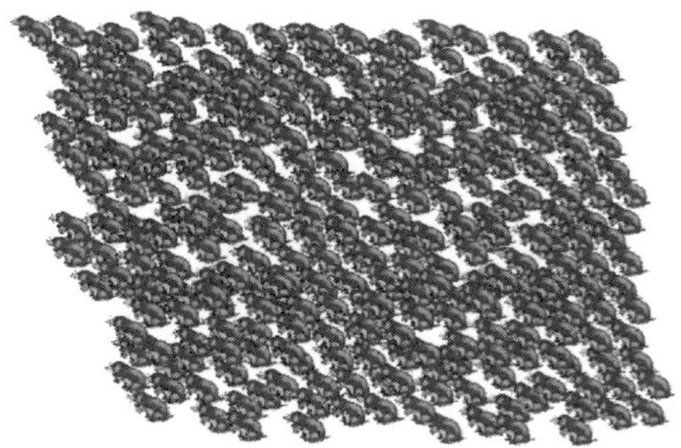

.......COMPLETE SOLUTIONS.......

1. Start with $P_4 + O_2 \rightarrow P_4O_{10}$

Balance the O $P_4 + \mathbf{5}O_2 \rightarrow P_4O_{10}$

2. Start with $IBr + NH_3 \rightarrow NI_3 + NH_4Br$

There are 3 H on the left and
4 H on the right. Make them both 12.
Do you remember finding least common denominators in arithmetic?

$IBr + \mathbf{4}NH_3 \rightarrow NI_3 + \mathbf{3}NH_4Br$

Balance the Br $\mathbf{3}IBr + 4NH_3 \rightarrow NI_3 + 3NH_4Br$

3. Start with $KrF_2 + H_2O \rightarrow Kr + O_2 + HF$

Balance the F $KrF_2 + H_2O \rightarrow Kr + O_2 + \mathbf{2}HF$

Balance the O $KrF_2 + \mathbf{2}H_2O \rightarrow Kr + O_2 + 2HF$

Balance the H $KrF_2 + 2H_2O \rightarrow Kr + O_2 + \mathbf{4}HF$

Balance the F (again) $\mathbf{2}KrF_2 + 2H_2O \rightarrow Kr + O_2 + 4HF$

Balance the Kr $2KrF_2 + 2H_2O \rightarrow \mathbf{2}Kr + O_2 + 4HF$

4. Start with $Li_3N \rightarrow Li + N_2$

Balance the N $\mathbf{2}Li_3N \rightarrow Li + N_2$

Balance the Li $2Li_3N \rightarrow \mathbf{6}Li + N_2$

5. Start with $BCl_3 + H_2O \rightarrow H_3BO_3 + HCl$

Balance the Cl $BCl_3 + H_2O \rightarrow H_3BO_3 + \mathbf{3}HCl$

Balance the H $BCl_3 + \mathbf{3}H_2O \rightarrow H_3BO_3 + 3HCl$

6. Start with $Al + HCl \rightarrow AlCl_3 + H_2$

Balance the H $Al + \mathbf{2}HCl \rightarrow AlCl_3 + H_2$

Balance the Cl $Al + \mathbf{6}HCl \rightarrow \mathbf{2}AlCl_3 + H_2$

Balance the H (again) $Al + 6HCl \rightarrow 2AlCl_3 + \mathbf{3}H_2$

Balance the Al $2Al + 6HCl \rightarrow 2AlCl_3 + 3H_2$

Chapter Twenty-three
Atomic Dating

Sometimes Joe wasn't paying full attention to his teachers.* That only happened on days ending in *y* or in months with less than 40 days in them. Between snacks he liked to turn to Darlene and tell her jokes. This was bad for three reasons: (1) It disturbed Darlene. She needed to pay attention, because she knew that if and when she married Joe, she would probably have to be the one earning an income; (2) Joe never spoke quietly, and this disturbed all the students around him when he talked to Darlene; (3) Joe's jokes were never really funny.

At this point in Fred's chemistry lecture, Joe had just finished a lollipop. He had chewed it rather than sucked on it. He turned to Darlene and read aloud the chart that Fred had written on the board

methane	CH_4
ethane	C_2H_6
propane	C_3H_8
butane	C_4H_{10}
pentane	C_5H_{12}

Joe asked her, "I bet you don't know what $C_{100}H_{202}$ is called."

In a quiet voice, she answered, "No, dear. What is $C_{100}H_{202}$ called?"

"Insane." Joe laughed loudly at his own joke.

"And what do you call," he continued, "a man who has been balancing chemical equations for 70 years?"

In a quiet voice, she answered, "No, dear. What do you call him?"

"Old. Ha ha ha ha ha."

Darlene just smiled and turned back to look at Fred.

Suddenly, she thought, "How did Joe know that it was $C_{100}H_{202}$ and not $C_{100}H_{201}$ or $C_{100}H_{203}$? Could he really have seen the pattern in CH_4, C_2H_6, C_3H_8, C_4H_{10}, C_5H_{12}? This would be so un-Joelike. He couldn't have noticed that the subscript on the H is found by doubling the subscript

---

* For those who have read some of the other Life of Fred books, they recognize this as an example of litotes—saying the opposite and attaching a *not*. Examples: Fred doesn't have a small nose. Joe isn't extremely bright. The surface of the sun isn't everybody's favorite vacation spot. (LIE-teh-tease)

on the C and adding 2. C_nH_{2n+2}" She turned to look at the paper he had been scribbling on.

Joe hadn't seen that pattern. Instead he had just written out the structural formula with a hundred Cs and had counted all the Hs.

When Joe went fishing, which was one of his favorite activities, he would count the scales on each fish he caught.

Darlene would often daydream about the day she would walk down the church aisle in her white silk dress with a thousand sequins and see Joe standing there eagerly waiting for her.*

Fred continued, despite the distractions that Joe was creating, "There are different ways that atoms couple together to make molecules."

The words *couple together* had Darlene's full attention. She elbowed Joe indicating that he should also pay attention. Darlene hoped that atoms getting together would suggest to Joe that he should someday propose to Darlene.

"The electrons that circle the nucleus of an atom arrange themselves in shells. Hydrogen and helium have only one shell. Hydrogen has one electron in that shell and helium has two. With its shell all filled up, helium is a very happy element—one of the noble gases. It doesn't want to interact with anything.

"When you get to larger elements with atomic numbers [number of protons] greater than two, then the electrons have more shells. The innermost shell is filled first—with two electrons. All these shells have fancy names, but the names are not important right now."

Joe wrote in his notes eggshell, seashell, taco shell. If Fred wasn't going to supply the names, Joe would.

Fred took a large onion and cut it in half and explained that the shells of electrons could be thought of as layers of an onion.

---

* If this day ever comes, he will probably be standing there eating a peanut-butter-and-jelly sandwich and counting the steps she makes.

Joe began to cry.

Darlene couldn't figure it out. That onion was ten feet away from Joe. It wasn't close enough to affect his eyes. "Why are you crying?"

Joe sniffed. "I could have used that onion on the hamburger I'm eating right now."

Fred overhead what Joe said. In fact, everyone in the room heard Joe. He tossed the onion to Joe and Joe smiled.

Fred continued, "Usually, it's only the electrons in the outermost shell that atoms are willing to share. The electrons in the shells closer to the nucleus [of protons and neutrons] are held so tightly, they are not available for any interaction. Those are called **core electrons**.

"The electrons in the outermost shell are called **valence electrons**. And, except for hydrogen and helium, eight is the happy number for valence electrons. Each of the noble gases after helium has a full set of eight valence electrons. [Neon, argon, krypton, xenon, and radon]

"When it comes to making chemical compounds, the rule is very simple: *Everyone wants to be like a noble gas*."

Joe wrote Eight is grate.

Darlene reached over and did a little editing Eight is ~~grate~~ great.

Fred pointed to the right side of the periodic table.

1	2	3	4	5	6	7	8	9	10	11	12	13	14	15	16	17	18
H																	He
Li	Be														O	F	Ne
Na	Mg														S	Cl	Ar
K	Ca	Sc	Ti	V	Cr	Mn	Fe	Co	Ni	Cu	Zn	Ga	Ge	As	Se	Br	Kr
Rb	Sr	Y	Zr	Nb	Mo	Tc	Ru	Rh	Pd	Ag	Cd	In	Sn	Sb	Te	I	Xe
Cs	Ba														Po	At	Rn
Fr	Ra																

"The group 17 elements, the halogens, all have seven valence electrons. Each of them would love to move over one space to the right and be like the noble gases with eight valance electrons.

"The Lewis structures for halogens all have seven dots. $:\underset{\cdot\cdot}{\ddot{\mathrm{F}}}\cdot$

"The group 1 elements, the alkali metals, [lithium, sodium, potassium, . . .] all have a single valance electron that they easily give up. Once it was gone, their electron shells would look like the noble gases with 2 in the first shell and 8 in every other shell."

Fred wrote the Lewis structure for sodium on the board. Na˙

Darlene nudged Joe. "Do you see that? Sodium and fluorine were made for each other!"

Her nudge was a little too enthusiastic. The top ball of his three-scoop ice cream cone went rolling across the floor.

Joe quickly ate the other two scoops before Darlene could bump him again. He kicked the scoop that was on the floor. It made a wet pink line as it rolled toward Fred.

"The kind of chemical bond between atoms makes all the difference. At one extreme is the **ionic bond**—for example, the bond between an alkali metal and a halogen. Sodium, Na ˙ , gives up is single valence electron to the halogen, such as chlorine, :C̈l. .

"The sodium atom becomes a sodium ion Na^+, and the chlorine atom becomes a chlorine ion Cl^-. Chlorine owns that extra electron *completely*."

Darlene wrote in her notes *Bonding makes all the difference* and showed that to Joe.

"Compounds with ionic bonds are hard, brittle, and melt at high temperatures. At room temperature, they are all solids.

"When chlorine rips sodium's valence electron off, the reaction is very exothermic. You can expect heat and maybe even sparks."

Darlene wrote something about a warm, passionate embrace, but gave it to Joe before we could see what she wrote.

"The Na^+ and the Cl^- ions stick with each other by electromagnetic attraction. That white stuff you put on french fries are not molecules of NaCl, but zillions of sodium and chlorine ions all nestled together in a clump.

"In water, those ions all separate and swim around separately. Pure water does not conduct electricity. But if you add salt, then the electricity can be ferried through on the ions."

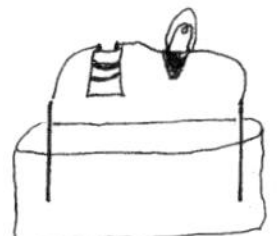

Fred hooked up a battery and lightbulb and stuck the ends in water. The light didn't light. When he added some salt, you know what happened.

Then Fred then did something that Bob Bunsen never did. He illustrated ionic bonding with a cartoon. Since Joe liked fishing, Fred drew the story of Sailor Sodium putting his single egg on a hook.

Your Turn to Play for Art Majors

Some people can draw better than Fred. (understatement) Fred's comic strip had *fishing* as the theme.

1. On a sheet of paper create a comic strip for ionic bonding that has *robbery* as its theme. One of the characters is Fearsome Fluorine who demands, "Gimme that pearl [electron] or else!" Sweet Sally Sodium responds, "But sir, this is the last pearl that I have!"

2. Draw a three-frame strip with the theme of *romance*. In the first frame, the famous knight, Sir Sodium, will declare, "I offer you my love, my last valence electron." In the second frame, Princess Florrie responds, "You are so nice. You make my shell complete." In the third frame they ride off together bonded by electrical attraction.

Fred did another piece of "art" in order to complete his discussion of ionic bonding.

Fred didn't like the fact that on the periodic table, those halogens which lacked an electron to achieve "noble gas happiness" were a lot closer than the alkali metals who needed to get rid of an electron to achieve complete shells.

He took a large paper periodic table and wrapped it into a cylinder and pasted the ends together.

O / F / He ‖ H \ Be
S / Cl / Ne ‖ Li \ Mg
Ar ‖ Na

Now the noble gases were at the center—as they should be. And flanking them on either side were the elements that were within one electron of achieving bliss.

Fred had created the three-dimensional merry-go-round periodic table of the elements that cannot be included in any flat chemistry book.

The O needs two electrons to get to Ne.

The Be needs to lose two electrons to get to Ne.

The S needs two electrons to get to Ar.

The Mg needs to lose two electrons to get to Ar.

Someday Fred may receive the Nobel prize in chemistry for his creation. You never know.

Chapter Twenty-four
Brotherly Love

Fred explained that the opposite of ionic bonding in which one of the two elements grabs the electron all for itself is **covalent bonding**. With covalent bonding, *complete* sharing occurs.

This covalent bonding is rare. It only occurs seven times in all of chemistry: H_2, N_2, O_2, F_2, Cl_2, Br_2, and I_2. The two atoms must be the same.

In the covalent bonding of hydrogen, for example, the two brothers equally share their two electrons. Somehow, they each feel that they have two electrons and their shells are complete.

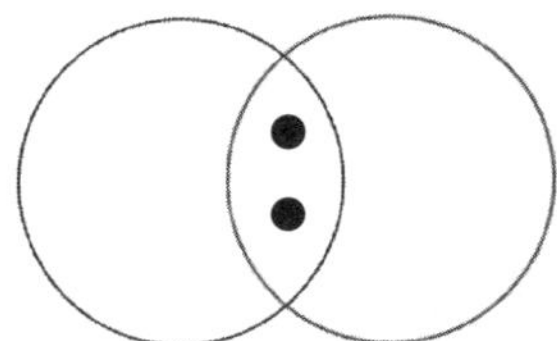

Why they both think they have two electrons is a real mystery. But they both think they are happy. Let's not argue with them.

When you get to your next course in chemistry, you will be told that electrons aren't little black balls that fly around in precise orbits. This is heartbreaking for some students.

They don't realize how often this occurs in education.

IN MATH . . .

✓ Your first grade teacher told you that you can't subtract 7 from 5.

Your beginning algebra teacher tells you that $5 - 7 = -2$.

✓ Your beginning algebra teacher told you that you can't take the square root of a negative number.

Your advanced algebra teach tells you $\sqrt{-1} = i$.

✓ Your advanced algebra teacher told you that you can't count the total number of natural numbers $\{1, 2, 3, 4, 5, \ldots\}$.

Your upper division college math teacher will tell you there are $\aleph_0$ natural numbers.

IN PSYCHOLOGY. . .

✓ In your beginning psychology class, you were told that all human behavior is only motivated by rewards and punishments.

Your upper division psych teacher tells you that we respond to more than just rewards and punishments.

IN SOME HOMES . . .

✓ They told you that Easter bunnies lay delicious chocolate eggs.

Later you learned that your pet bunny's "eggs" didn't taste very good.

IN PHYSICS. . .

✓ Your high school teacher told you that the universe is entirely composed of matter and energy.

In college physics your teacher tells you that matter and energy are a small fraction of the stuff of the universe. The teacher says that dark matter is much more common than regular matter and then tells you no one knows what it is.

In later chem courses, electrons will not be little black balls. They will have no color and they will be clouds. An electron could be virtually anywhere with probabilities assigned to each place in space.

Two hydrogen atoms sharing two electrons might look like this:

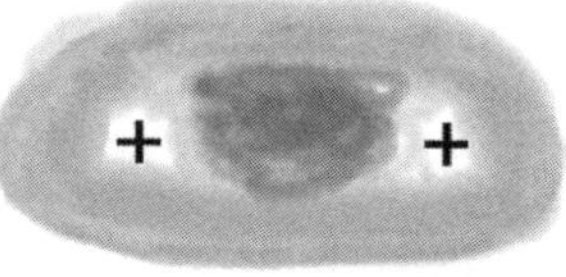

Wait a minute! I, your reader, need to say something. I bought this chem book and you tell me you are going to treat me like a baby.

What do you, my reader, have in mind?

Why don't we just skip the "little black ball" stuff for electrons. We can just move on to the more advanced approach. I can take it. I'm tough.

You mean that you want me . . . no I can't. We need to stay with picturing electrons as little balls circling the nucleus just like the planets circle the sun. You really aren't ready for the electron cloud approach.

I insist! I'm ready. I know all the elementary stuff. You taught it to me. I know that hydrogen has one proton in the middle and one electron circles around the outside. Helium has two protons, two neutrons, and two electrons . . . except, maybe, for those isotope things.

I hate to do this.

Do it! Otherwise, I'll report you to the Kiddie Chem Cops for only teaching beginning chemistry.

I thought that the ~~spearmint~~ experiment that the students did to determine the empirical formula for silver chloride was pretty tough.

You are stalling. Show me how they describe where an electron will be when it is like a cloud rather than like a ball.

Okay. Here goes. You tell me when to stop.

You are working in three dimensions.

I knew that.

In beginning algebra you graphed things in two dimensions, and you had the X–Y axes. The points looked like this: (x, y).

In real life, things are in three dimensions—a little like looking at a corner of a room. To locate something in the room, like a fly, you have to give the x coordinate, the y coordinate, and the z coordinate.

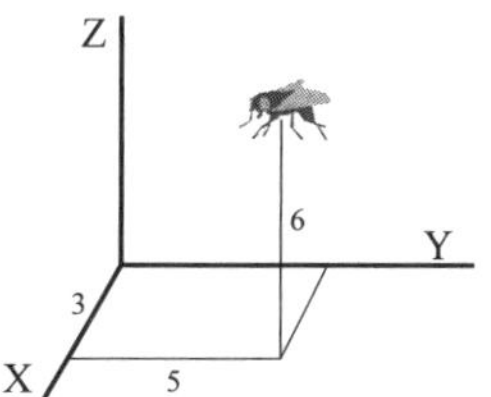

The fly is 3 in the x direction, 5 in the y, and 6 in the z. Its coordinates are (3, 5, 6).

I'm still here. If this is as bad as it gets, I'm a genius.

Okay. Let's look at a single electron around a single proton.

Hydrogen atom!

Yes. The probability that an electron is at a particular point is a function of where that point, (x, y, z), is. There is a very low probability that the electron is ten miles away from the proton.

Everybody knows that.

We will call that probability function $\psi(x, y, z)$. ψ is the Greek letter psi. It is pronounced sigh.

I got a question. Why don't they use f or g for the function name. I'm much more used to things like y = f(x). Why ψ? Do you like my poetry?

You are a multi-lingual poet! They use ψ, because it is the traditional letter used in this context. It's just like the fact that everyone in this country drives on the right side of the road. It's tradition.

Actually, I'm trying like crazy to keep this cloud description of an electron in a hydrogen atom as simple as possible. $\psi(x, y, z)$ is the wave function. The probability that an electron is at (x, y, z) is given by

$$\text{Probability} = |\psi(x, y, z)|^2.$$

But this is just a picky point. It's okay right now to think of the function ψ as a probability.

You will need an equation to compute the chances that the electron in a hydrogen atom is at the point (x, y, z).

Yeah. Gimme the equation. That's what I've been waiting for.

Okay. Here it is.

$$\frac{-h^2}{2m}\left(\frac{\partial^2\, \psi(x, y, z)}{\partial x^2} + \frac{\partial^2\, \psi(x, y, z)}{\partial y^2} + \frac{\partial^2\, \psi(x, y, z)}{\partial z^2}\right) = \left(E + \frac{e^2}{r}\right)\psi(x, y, z)$$

[stunned silence]

Would you like me to continue? This is the simplest case—one electron. It is just a three-dimensional time-independent formulation of **Schrödinger's equation**. You will encounter partial derivatives like $\frac{\partial f}{\partial x}$ in the later part of the second year of calculus. $\frac{\partial^2 f}{\partial x^2}$ has nothing to do with squaring. It is the partial derivative of the partial derivative of function f (where f has at least two independent variables).

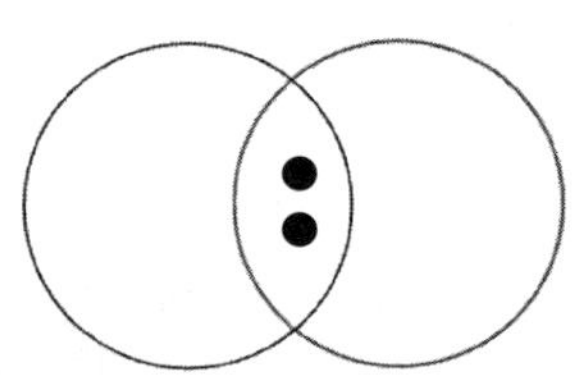

I have suddenly taken a deep love of the picture of two hydrogen atoms equally sharing their two little electrons. Those little spheres are so cute.

Covalent bonding you called it. That's nice.

There was a reason why your first grade teacher told you that you can't subtract 7 from 5.

Your Turn to Play

1. Balance this skeleton equation: $Al + HCl \rightarrow AlCl_3 + H_2$

A small description of what that equation represents: A piece of aluminum is dropped into hydrochloric acid. $AlCl_3$ is a precipitate and hydrogen gas bubbles out of the acid. An explosion may occur if someone is standing nearby and smoking; the hydrogen and the oxygen (from the air) would combine exothermically.

2. Magnesium sulfate, $MgSO_4$, is another name for Epsom salts.

Atomic weights: Mg = 24.31
S = 32.01
O = 16.00

Five pounds of Epsom salts contains how many pounds of magnesium?

3. Everyone knows that oxygen is one of the seven elements that can form diatomic, covalent molecules. (I said that on the fifth line of this chapter, but I didn't use all those fancy words.)

When two hydrogens get together they share a bonding pair. H : H When two chlorine atoms get together they also share a bonding pair of electrons.

$$:\ddot{\underset{..}{Cl}}:\ddot{\underset{..}{Cl}}:$$

But the Lewis structure for a single oxygen atom is $:\ddot{\underset{..}{O}}$

Oxygen only has six valence electrons. That's why it likes to combine with two hydrogen atoms.

$$\begin{array}{c} H \\ :\ddot{\underset{..}{O}}:H \end{array}$$

Sometimes the sharing of a pair of electrons is written as "—" instead of using dots.

$$\begin{array}{l} H \\ | \\ O - H \end{array}$$

How do two oxygen atoms manage to find happiness with each other? We haven't covered this in the text. I'm asking you to think about it and make a wild guess.

.......COMPLETE SOLUTIONS.......

1. Start with $Al + HCl \rightarrow AlCl_3 + H_2$
Balance the H $Al + \mathbf{2}HCl \rightarrow AlCl_3 + H_2$
Balance the Cl
(make them both 6) $Al + \mathbf{6}HCl \rightarrow \mathbf{2}AlCl_3 + H_2$
Balance the H (again) $Al + 6HCl \rightarrow 2AlCl_3 + \mathbf{3}H_2$
Balance the Al $\mathbf{2}Al + 6HCl \rightarrow 2AlCl_3 + 3H_2$
(You might have done the balancing in a different order.)

Balanced. 2 Al on each side; 6 H on each side; 6 Cl on each side.

2. From the *Memory Aids* (p. 137): When you are working with weights, convert things to moles.

One mole of $MgSO_4$ contains one mole of Mg.

The molecular weight of $MgSO_4$ is 120.32. (The math: 24.31 + 32.01 + 4×16.00.)

The Mg in $MgSO_4$ is 20.20% by weight. (The math: 24.31 ÷ 120.32)
20.20% of 5 lbs. is 1.010 lbs.

This could also have been done using conversion factors, but it takes longer. 1 pound ≈ 453.6 grams

$$\frac{5\text{ lbs.}}{1} \times \frac{453.6\text{ g}}{1\text{ lb.}} \times \frac{1\text{ mole }MgSO_4}{120.32\text{ g}} \approx 18.85\text{ moles of }MgSO_4$$

Since 18.85 moles of $MgSO_4$ contains 18.85 moles of Mg, we need to convert 18.85 moles of Mg into pounds.

$$\frac{18.85\text{ moles Mg}}{1} \times \frac{24.31\text{ g}}{1\text{ mole Mg}} \times \frac{1\text{ lb.}}{453.6\text{ g}} \doteq 1.010\text{ lbs.}$$

3. Instead of each atom sharing just one electron (as H and Cl did), they will each share two electrons.

$$\ddot{\underset{\cdot\cdot}{O}} :: \ddot{\underset{\cdot\cdot}{O}} \quad \text{or} \quad \ddot{\underset{\cdot\cdot}{O}} = \ddot{\underset{\cdot\cdot}{O}}$$

It's called a **double bond**.

Chapter Twenty-five
World's Most Popular Bonding

Fred knew he had some explaining to do. There are six alkali metals (Li, Na, K, Rb, Cs, and Fr) and five halogens (F, Cl, Br, I, and At), so there are certainly less than 40 pairings that are ionic—where one atom takes the extra electron and no sharing happens.

There are seven "brotherly love" covalent bonds in which the electrons are completely evenly shared (H_2, N_2, O_2, F_2, Cl_2, Br_2, and I_2).

Adding those two kinds of chemical bondings, we have less than 50 compounds accounted for. But there are a zillion different kinds of molecules.

By a quick calculation, that leaves 99.9999999999999% of all chemical compounds that haven't been talked about.

All the hands in the classroom went up, except Joe's. He was busy stirring his hot chocolate. The students all noticed that Fred had left out a large class of chemical bonds.

If ionic bonding means that one of the atoms steals an electron completely away . . .

If covalent bonding means that the pair of atoms equally share electrons . . .

. . . then an unequal sharing of electrons is all that's left. It's called polar covalent bonding.

Hey! I like that type font. I have never used it before. Do you think anyone would object if I used it for the rest of this book?

I, your reader, would! Please go back to your regular Times New Roman font.

Okay. But I do like this font.

Knock it off! Get back to explaining polar covalent bonding.

What's there to say? Two or more atoms meet and they have a tug of war over the valence electrons. Some atoms pull harder than others. The noble gases don't pull at all. They are happy with the electrons they have. All their shells are already filled.

How hard an element pulls electrons is call its **electronegativity**. Leave it to chemists to invent an eight-syllable word. I would have called it pull-power.

Here's an abbreviated periodic table with electronegativities to give you an idea who has the pull-power.

		H 2.2				

group ➟

1	2			15	16	17
Li 1.0	Be 1.6			N 3.0	O 3.4	F 4.0
Na 0.93	Mg 1.3			P 2.2	S 2.6	Cl 3.2
K 0.82	Ca 1.3					Br 3.0
Rb 0.82						
Cs 0.79						

Some notes . . .

♪#1: I could have typed out the whole periodic table, but what would be the point? You get the idea from these entries. The halogens in period 17 can really pull electrons off any atoms they meet. The alkali metals in period 1 are the most generous, kindest elements you would ever meet. They would almost happily give you the shirt (electron) on their back if you wanted it.

♪#2: Recall that encounters between period 1 and period 17 atoms result in ionic bonding. If a fluorine atom (upper right corner) ever bumped into a cesium (Cs) atom (lower left corner), the fluorine atom wouldn't even say, "Please" before removing cesium's lone electron from his outer shell.

♪#3: Darlene wrote in her notes: *I have met the man of my dreams (Joe). I want to Cs.*

♪#4: Because this is chemistry and not mathematics, everything is more in shades of gray rather than in black and white. In math we don't say that two plus two is approximately equal to four. We don't say that the base angles of an isosceles triangle are nearly congruent.

Chemistry is more like real life. Few people can sing the really high notes. Few can sing the really low notes. Between the highest sopranos and the lowest basses, are 99% of the people in choirs.

Between the few cases of true ionic bonding (such as NaCl) and the few cases of pure covalent bonding (the famous seven: H_2, N_2, O_2, F_2, Cl_2, Br_2, and I_2) we have the billions of compounds (the polar covalent ones) in which the electrons are unevenly shared.

If you want to know how unequally the sharing is, just look at their respective electronegativities.

♪#5: What happens when you don't share the electrons equally? Part of the molecule is more electrically negative and part is more positive—a little like north and south ends of a magnet. That's why they call it *polar* covalent bonding.

A water molecule isn't straight like this: H — O — H. It is bent.*

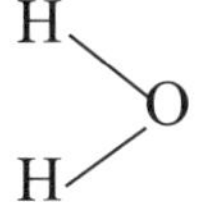

The "O" end (electronegativity = 3.4) has more of the electrons than the "H" ends (electronegativity = 2.2).

When salt (NaCl) dissolves in water, the Na^+ ions tend to gather around the "O" end of water molecules.

♪#6: Because H_2O is a polar molecule, molecules of water are attracted to other molecules of water. Because of that, its harder to change water from a liquid into a gas. (Changing into a gas involves pulling the molecules apart from each other.) Because of that, the oceans and lakes are not empty. Because of that, you can get a slurp of water to drink. Because of that, you are alive.

In short, water is a polar molecule means you are alive.

---

* If you want to impress your friends at a party, you can tell them that the angle is 104.5° (which it is).

♪#7: Carbon dioxide (the bubbles in soft drinks) has polar covalent bonds (Doesn't almost everybody?).

Electronegativity of C = 2.6.

Electronegativity of O = 3.4.

In CO_2, the two oxygen atoms tend to hold on to the electrons more than the carbon atom.

Is the molecule polar? Nope.

The molecule is in a straight line. O — C — O

The bonds are polar; the molecule isn't.

♪#8: The **dipole moment** is the measure how polar a molecule is. The dipole moment of CO_2 is 0. The dipole moment of H_2O is 1.85. That means you are alive. (See ♪#6.)

♪#9: Stuff that is near the ionic bonding end of the spectrum (big differences between the electronegativities of the atoms) tends to be brittle, hard, and have a high melting point.

Stuff that is near the covalent bonding end tends to be flexible and have a low melting point. Plastic bags—flexible and melt easily if you put them on a hot electric kitchen stove*—are a good example.

♪#10: Stuff that dissolves in water is called **soluble**. Stuff that doesn't is called insoluble. (← for Joe's benefit) Most things are somewhere between totally soluble and totally insoluble.

♪#11: Stuff that makes a lot of ions when dissolved in water (like NaCl) is called an **electrolyte**. It will conduct electricity. Stuff that doesn't make lots of ions when dissolved in water (like sugar) is called a nonelectrolyte. (← for Joe's benefit) This is chem. Lots of gray. Virtually everything makes some ions when in contact with water.

Tall-short/soluble-insoluble/heavy-light/electrolyte-nonelectrolyte/smart-dull/pretty-plain/rich-poor/ionic-covalent/dark-light . . . in all of life very few are at the extremes. 99.9999% are somewhere in the middle.

None have the right to brag. None need be ashamed. It's just like chemistry—almost everybody and everything is in between.

Even in math, I never heard a number brag, "I'm the largest."

---

* Don't you dare! Really. This is not a RCE (Recommended Chem Experiment). That will make a melted mess. It might catch fire.

Please write your answers to the *Your Turn to Play* before you turn the page and look at my answers. Please.

Your Turn to Play

1. Name some things that you are glad are insoluble.
2. Joe eats so much sugar that his mother once said that he sweats sugar water rather than salty water like everyone else. She might have been exaggerating.

Joe grabs the two wires, one with each hand.

Darlene, who has just washed her hands, grabs the two wires with her wet hands.

Fred, who realizes the danger and is sweating like crazy, is handed the two wires.

Which of these three will get the greatest shock?

3. There are three acids typically found in high school and college chem labs:

HCl	hydrochloric acid
H_2SO_4	sulfuric acid
HNO_3	nitric acid.

They all sit there in nice glass bottles waiting to be used in experiments.

My favorite acid is hydrofluoric acid, HF. This stuff is so BAD, it will eat through glass (silicon dioxide).

The skeleton equation is $HF + SiO_2 \rightarrow SiF_4 + H_2O$

Silicon tetrafluoride (SiF_4) is a gas that you can see bubbling up through the solution as the hydrofluoric acid eats away at the glass.

Balance the skeleton equation.

.......COMPLETE SOLUTIONS.......

1. Your skin. Buildings. Beaches. (Your list may be different than mine.) Imagine a sugar airplane in the rain.

"I'm so glad my fish are insoluble."

2. Sugar is a nonelectrolyte. Joe is in less danger.

Water doesn't conduct electricity. Darlene is in less danger.

Fred is going to receive the biggest shock. NaCl is a great electrolyte.

3. The skeleton equation $HF + SiO_2 \rightarrow SiF_4 + H_2O$

It's not clear which element to balance first. The Si is already balanced.

Balance the H $\mathbf{2}\,HF + SiO_2 \rightarrow SiF_4 + H_2O$

Balance the O $2\,HF + SiO_2 \rightarrow SiF_4 + \mathbf{2}\,H_2O$

Balance the H (again) $\mathbf{4}\,HF + SiO_2 \rightarrow SiF_4 + 2\,H_2O$

Balanced! 4 H on each side; 1 Si on each side; 2 O on each side.

English lesson . . .

In the equation $4\,HF + SiO_2 \rightarrow SiF_4 + 2\,H_2O$

the HF and SiO_2 are called the **reactants**.

The SiF_4 and H_2O are called the **products**.

reactants products

exercise + good diet → health + long life

Chapter Twenty-six
Ionic Destiny

Fred liked to keep things simple. That wasn't always possible, but if it was possible, he would teach it that way. With chemical bonding, the central idea was how those valence electrons would be shared. At one extreme, one atom grabbed the electron all for itself—ionic. At the other extreme, there was perfect sharing—covalent. In between were all the examples of unequal sharing—polar covalent.

It was really nice that the group 1 elements all were willing to give up exactly one electron and form the ions H^+, Li^+, Na^+, K^+, Rb^+, and Cs^+. The group 2 elements, beryllium, magnesium, calcium, strontium, barium, and radium, all liked to give up two electrons and form the ions Be^{2+}, Mg^{2+}, Ca^{2+}, etc.

Over on the right side of the periodic table, things were equally wonderful.

group ➡ 1	2			16	17	18
H^+				O^{2-}	F^-	He
Li^+				S^{2-}	Cl^-	Ne
Na^+	Mg^{2+}				Br^-	Ar
K^+	Ca^{2+}				I^-	Kr
Rb^+	Sr^{2+}					Xe
Cs^+	Ba^{2+}					Rn

To make things more complicated, chemists call the positive ions **cations** (pronounced CAT-ions) and the negative ions **anions**.

I can't figure that out. Is saying, "cation" that much shorter than saying, "positive ion"? Maybe this is their way of apologizing for using *electronegativity* instead of *pull power*.

If you like cats—have positive feelings toward cats—then it's easy to remember cations are the positive ions.

Fred was almost apologetic when he had to break the bad news, "In the middle of the periodic table are a bunch of metals that don't have a single ionic destiny. Take chromium (Cr) for example. Some days it will form compounds by giving up two electrons: Cr^{2+}. And some days, it will give up three: Cr^{3+}. It seems to be happy in either state."

Fred filled in several of the inner groups.

1	2	6	7	8	9	10	11	12	14	16	17	18
H^+										O^{2-}	F^-	He
Li^+										S^{2-}	Cl^-	Ne
Na^+	Mg^{2+}										Br^-	Ar
K^+	Ca^{2+}	Cr^{2+} Cr^{3+}	Mn^{2+} Mn^{3+}	Fe^{2+} Fe^{3+}	Co^{2+} Co^{3+}	Ni^{2+} Ni^{3+}	Cu^+ Cu^{2+}	Zn^{2+}			I^-	Kr
Rb^+	Sr^{2+}						Ag^+	Cd^{2+}	Sn^{2+} Sn^{4+}			Xe
Cs^+	Ba^{2+}						Au^+ Au^{3+}	Hg^{2+} Hg_2^{2+}	Pb^{2+} Pb^{4+}			Rn

"In groups 6–10, things look sort of nice. All of those will lose either two or three electrons. Then in group 11, things get really weird. Copper (Cu) will lose only one or two. And silver (Ag), only one. And gold (Au), one or three.

"Mercury is in a case by itself. When it loses two electrons (Hg^{2+}) it looks very much like its friends in the same group 12—zinc (Zn^{2+}) and cadmium (Cd^{2+}). But two mercury atoms can get together and each lose one electron (Hg_2^{2+}).

Wait a minute! If none of those students sitting there are going to ask Fred the obvious question, I, your reader, need to ask it.

The obvious question?

Yes. The obvious question: Why? Why do things get so goofy in the middle of the periodic table?

When I was little and asked my mother a hard question, she would often answer, "Because." That kept my young brain from breaking into a thousand pieces.

But you are not my mother! You are the author. I'm asking a simple question: How come there is such an ionic mess in the middle of the table? A simple question deserves a simple answer.*

Have you forgotten that question you asked me two chapters ago when you wanted me to move from the little-black-ball image of an electron to the cloud image?

You mean this explanation is going to end up with another equation like the Schrödinger's equation?

$$\frac{-h^2}{2m}\left(\frac{\partial^2 \psi(x, y, z)}{\partial x^2} + \frac{\partial^2 \psi(x, y, z)}{\partial y^2} + \frac{\partial^2 \psi(x, y, z)}{\partial z^2}\right) = \left(E + \frac{e^2}{r}\right)\psi(x, y, z)$$

It won't involve any equations to explain the ionic mess. I just don't want to break your brain into a thousand pieces. Can't I just say, "Because"?

No. I'm tougher now. Explain.

Okay. Here goes. I'm going to stick a vertical strip on the right side of this story so that if your baby brothers or sisters accidentally open this book, they can easily skip over this section if they want to.

Once upon a time, (I'm keeping things as simple as possible), we used to have orbits that the little electron balls would roll around in. We have since learned that electrons are not in any particular spot. Their locations can only be described as probabilities. The probability that an electron is at (x, y, z) is given by $|\psi(x, y, z)|^2$ where ψ is the function in the *general* Schrödinger's equation—not the simplified equation (given earlier on this page) that is just for the hydrogen atom.

I promised that I wouldn't give you any equations in this story.

So instead of *orbits*, we call these probabilistic clouds **orbitals**. To describe which orbital an electron is in requires four

---

* That's silly. In Real Life, some of the simplest questions are the ones that our best thinkers have wrestled with and have had to shrug their shoulders.

pieces of information. That is not so weird. If I want to send you a letter, I have to know your country (USA)(one item), your state (Kansas)(second item), your city (Topeka)(third item), and your street address (123 Main Street)(fourth item).

The location of any particular electron (cloud) is given by the four numbers n, ℓ, m_ℓ, and m_s.

Let's start with n. It is called the **principal quantum number**. Every chem book and every chemist uses the letter n for the principal quantum number. It's a tradition.

Electrons, which are negative, love to be close to the nucleus, which has protons with positive charges. The shell or energy level closest to the nucleus is called n = 1. The next closest is n = 2. Then n = 3, etc. Chemists can sometimes be very logical.

Shell is the traditional word, but most shells that I know of are hard and have definite surfaces. Chemists' shells are fluffy like cotton candy.

In a hydrogen atom, the electron loves to be in the n = 1 energy state. You can push that electron into shells farther away from the nucleus by adding energy (heat it or zap it with electricity), but the minute you turn your back, the electron will flop back down to n = 1 releasing that energy in the form of light.

Electrons that fall down to the n = 1 energy level release ultraviolet light, which human eyes can't see.

Electrons that fall down to the n = 2 shell release visible light.

Electrons that fall down to the n = 3 energy level release infrared light, which we experience as warmth.

So take neon. It has all of its n =1 and n = 2 shells filled. Stick that gas in a tube and run an electric charge through it. The electrons get excited and hop up into the n = 3, n = 4, or n = 5 shells and then flop down to the n = 2 shell. Visible light is produced.

Starring Fred Gauss

Someday Fred's name will be up in neon lights.

Skipping down from, say, n = 5 to n = 2 is like going down stairs. You can't go down 1¾ stairs. You can't drop from n = 5 to n = 3¼. Only particular wavelengths (colors) of light will be generated when electrons are bouncing down the stairs.

Secondly, the distance from, say, n = 5 to n = 2 will vary depending on how many protons are in the nucleus. The single proton in the hydrogen nucleus will attract electrons differently than mercury with 80 protons. Thinking of the principal quantum numbers (n) as stairs, the height of the steps will be different for H than for Hg.

Putting those two facts together, each element will generate its own particular set of colors (wavelengths). Each element has its own fingerprint.

Hey! All this stuff is nice. I, your reader, never knew about neon lights before . . . but you haven't answered my question yet about why iron has two cations (Fe^{2+} and Fe^{3+}).

I'm getting there. Those "simple questions" containing *why* usually take the longest to answer.*

When you pass white light through a triangularly shaped piece of glass (known as a prism) it splits the light into a rainbow. White light consists of all the colors.

When you pass the light generated by a neon light through a prism you get very specific lines of color with black between those lines. You get a **line spectrum**.

With hydrogen you will be able to see four specific lines of color: red, blue-green, blue, and indigo. The red will be a very

---

* When your kid asks you why apples always fall downwards and you answer, "It's because of gravity," have you really said anything? You have named the mystery with the word *gravity*, but just putting a name on something doesn't explain anything. You might as well have said, "Things fall because the earth sucks." Or you might have said, "Things fall toward your feet so that we can know which way *down* is."

The truth is that most adults don't have the foggiest idea why things fall downward.

The bravest parents offer the true answer, "I don't know."

particular shade of red. It's wavelength will be 656.3 billionths of a meter (known as 656.3 nanometers). Colors are waves. Reds come in shades between 622 and 770 nm in wavelength.

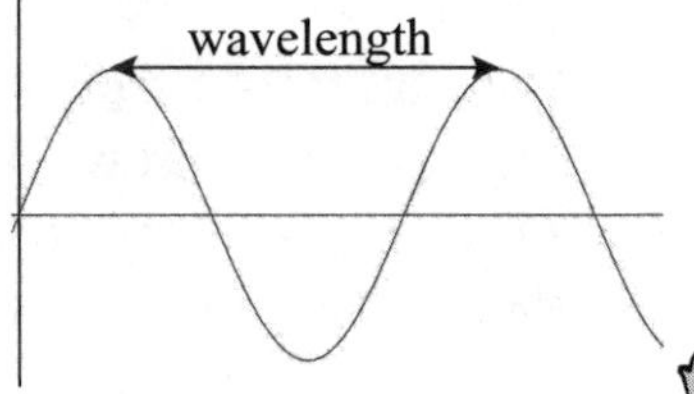

And there will be specific lines in the invisible parts of the spectrum in both the infrared and the ultraviolet regions. Each element has its own signature.

Now to get back to your question.

Thank you.

The four numbers in the address of an electron are n, ℓ, m_ℓ, and m_s. We have already looked at n, the principal quantum number.

The second number, ℓ, describes which subshell the electron is in. Depending on which chemist you are talking with, ℓ is called the **orbital quantum number** or the **angular-momentum quantum number** or the **azimuthal quantum number**.

The number of values of ℓ depends on the value of n. The math is pretty easy.

If n = 1, then there is 1 possible subshell.
If n = 2, then there are 2 possible subshells.
If n = 3, then there are 3 possible subshells.
If n = 4, then there are 4 possible subshells. etc.

Chemists are not mathematicians. The logical thing to name the four subshells of an electron in the fourth shell (n = 4) would be ℓ = 1, 2, 3, 4. Instead, they name those four subshells ℓ = 0, 1, 2, 3.

The two subshells of the n = 2 shell are ℓ = 0, 1.

Since that wasn't hard enough, ℓ = 0 is called sharp. ℓ = 1 is called principal. ℓ = 2 is called diffuse. ℓ = 3 is called fundamental. Those words were invented when chemists were first looking at spectra (plural of stectrum) and didn't know exactly what they were looking at. Often these four words are shortened to *s, p, d,* and *f.*

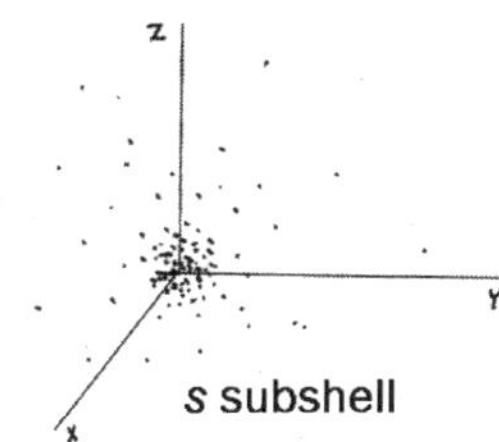

These values of ℓ describe the shape of the electron clouds in the subshells. *s* is nice. Electrons in the *s* subshell are in the shape of a sphere, except there is no boundary on the ball. The probability of finding an electron depends on only the distance from the origin.

When $\ell = 1$ (the *p* subshells) the electron cloud is in the shape of two teardrops that are kissing (officially called **lobes**).

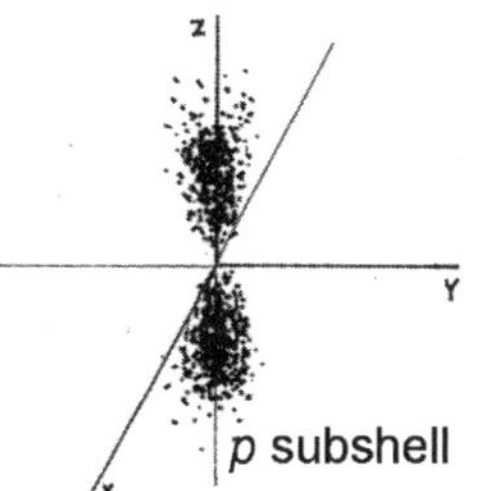

With $\ell = 2$ (the *d* subshells), things really start to get extreme. The *d* subshells (in four cases) will have four lobes, and in the fifth case it will look a *p* subshell (two lobes) with a doughnut around its middle.

Now we are getting close to explaining why manganese can have two cations (Mn^{2+} and Mn^{3+}).

But first, I need to take a little break.

What! You are getting close to answering my question, and you want a take a break! Please. Let's get on with the show.

How about we make a deal? I take my little break and there's no *Your Turn to Play* at the end of this chapter?

You got a deal.

In the periodic table the columns were called groups. There are 18 groups. The rows were called periods. There are seven periods.

	1	2	3	4	5	6	7	8	9	10	11	12	13	14	15	16	17	18
1	H																	He
2	Li	Be											B	C	N	O	F	Ne
3	Na	Mg											Al	Si	P	S	Cl	Ar
4	K	Ca	Sc	Ti	V	Cr	Mn	Fe	Co	Ni	Cu	Zn	Ga	Ge	As	Se	Br	Kr

5	Rb	Sr	Y	Zr	Nb	Mo	Tc	Ru	Rh	Pd	Ag	Cd	In	Sn	Sb	Te	I	Xe
6	Cs	Ba	La Ce Pr Nd Pm Sm Eu Gd Tb Dy Ho Er Tm Yb Lu Hf Ta W Re Os Ir Pt Au Hg Tl Pb Bi													Po	At	Rn
7	Fr	Ra																

My little break consists of mentioning that the first row (H and He) are elements with principal quantum number n = 1.

The second period are those with n = 2. And so on.

Now back to answering your question.

The electron in the hydrogen atom likes to be in the 1*s* (n = 1 and ℓ = 0) level, but give it a touch of energy (a spark, for example) and it can bounce up to 2*s*, or 3*p*, or 4*f*—up to any higher energy level and any subshell it likes.

All the n = 2 orbitals (2*s* and 2*p*) have the same energy level. All the n = 3 orbitals (3*s*, 3*p*, and 3*d*) have the same energy level.

For hydrogen: 2*s* = 2*p* in energy

3*s* = 3*p* = 3*d* in energy

4*s* = 4*p* = 4*d* = 4*f* in energy.

Now comes the fun part. When you have an element with lots of electrons, like manganese with 25 of them, all those electrons repel each other and mess up those pretty electron cloud shapes.

Those subshells get squished and contorted. The result? 2*p* will have a higher energy level than 2*s*. 3*d* will have a higher energy level than 3*p*.

When we get to n = 4, where all the action starts,

4	K^+	Ca^{2+}	Cr^{2+} Cr^{3+}	Mn^{2+} Mn^{3+}	Fe^{2+} Fe^{3+}	Co^{2+} Co^{3+}	Ni^{2+} Ni^{3+}	Cu^+ Cu^{2+}	Zn^{2+}

subshell 4*s* has less energy than 3*d* even though it is at a higher quantum number level. The electron clouds get so tangled that a poor little electron can't easily decide which of two orbitals has the lowest energy. On some days iron can be happy giving up two electrons (Fe^{2+}), and on other days it thinks that giving up three electrons will make it most satisfied.

That's why things get goofy in the middle of the periodic table.

Thank you.

Chapter Twenty-seven
Please Pass the Proton

Fred really didn't like explaining all that *s* and *p* orbital stuff. When iron combines with chlorine, chemists know that chlorine can only form one anion (Cl^-) and that iron will willingly donate either two (Fe^{2+}) or three electrons (Fe^{3+}). The result will be either $FeCl_2$ or $FeCl_3$, which they call iron(II) chloride or iron(III) chloride.

Older chemists call these compounds ferrous chloride and ferric chloride. The *ous* ending goes with the lower valence and the *ic* ending with the atom that has donated more electrons.

CuCl is cuprous chloride and $CuCl_2$ is cupric chloride.

Knowing the shape those dumbbell *p* orbitals—does that make a difference when you are doing experiments? So much poor teaching gets involved with the silly details. The students often miss the big picture.

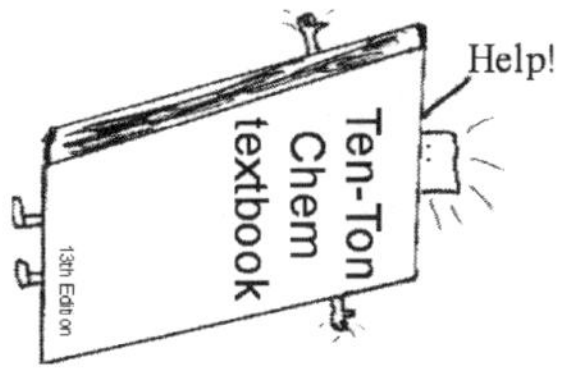

Some chem textbooks are 1200 pages long and cost more than \$200. One of them weighs 5 lbs. 7 oz.—the weight of some newborn babies. They dig down to the m_s level in the four-part address (n, ℓ, m_ℓ, and m_s) and give you Hund's Rule as the third section of the **aufbau principle** for the filling order for electron orbitals.*

---

* I'm not kidding. You may have to memorize this stuff for AP (college level) chem courses. And one student in a thousand might ever use it again—and then only if they have to teach college chemistry to the next unfortunate generation of chem students.

The aufbau principle sounds really elegant. It isn't. The word is stolen from the German der Aufbau, which means construction or organization. Why simply say "the construction rules for sticking electrons in their shells and subshells," when you can declare, "the aufbau principle and Hund's Rule for allocating electrons into orbitals and suborbitals as denoted by principal quantum number (n), angular-momentum quantum number (ℓ), **orbital magnetic quantum number** (m_ℓ), and **spin magnetic quantum number** (m_s)"? Gasp. Cough. Choke

Fred patiently waited during Chapter 26 while the author had answered the reader's question about why all those metals in period 4 had multiple oxidation states. Finally, Fred could get back to introducing one of his favorite parts of chemistry: acids & bases.

He began, "Acids have a taste we sometimes enjoy: lemons, grapefruit, vinegar, and bubbly soft drinks."

Joe held up an empty Sluice bottle. None of his 14 Sluice bottles that were arranged in a circle around him contained a drop. Joe felt very good that he was understanding chemistry. Actually, all he understood was the meaning of *bubbly soft drink.*

"The opposite of acids are bases," Fred continued. "Drain cleaners and antacids like Tums® are examples of bases.

"All of the chemical reactions we have studied so far in this lecture have involved electrons—outside the nucleus. We have looked at reactions in which one atom stole an electron completely from another atom—ionic bonding. We have looked at seven cases where two identical atoms completely shared electrons: H_2, N_2, O_2, F_2, Cl_2, Br_2, and I_2—covalent bonding. And we have looked at unequal sharing—polar covalent bonding.

"If we delve deeply into the atom beneath the electron clouds and into the nucleus, we are in the world of nuclear physics.* In nuclear physics, we smack an overweight atom—such as uranium, which has 92 protons—with a neutron. That breaks the nucleus into smaller nuclei (plural of nucleus) and sends out a bunch more neutrons. If you have a whole bunch of uranium atoms standing together in a crowd and you smack one of them, he will split apart and neutrons will smack other uranium atoms that are standing close by. Pretty soon the whole crowd has been smacked and split. This is called nuclear fission. If you keep the rate of splitting down by inserting non-uranium stuff that absorbs some of the neutrons that are flying around, then the uranium gets nice and warm and keeps an even "fire" burning. You boil water into steam, drive turbine generators with the steam and produce tons of electricity.

"If you don't insert the non-uranium stuff to absorb some of the neutrons flying around, then very quickly lots of the uranium atoms are getting smacked and you have ka-boom! The mass lost in the bomb over

---

* Which some chemists like to call "nuclear chemistry" so that they can get to study it also.

Hiroshima was the weight of a dime. That tiny mass was equal to forty million pounds of TNT. $E = mc^2$ (E = energy. m = mass. c = speed of light.)

"When you burn coal or set off dynamite, only the electrons are involved. *Nuclear* reactions release much more energy.

small essay

We Need Electricity

Electricity does more than run our computers, light our houses, and operate the pumps at the gas station. Turn off our electrical power plants and 90% of our population would be dead within a year.

On day one, your water is no longer running. In three days all the grocery stores are empty. In a month, virtually no cars and trucks are running. Could you grow enough in your backyard to survive? Imagine walking to the nearest lake with a couple of five-gallon buckets to get water for the carrots you planted.

So we need electrical power. The problem is that generating electricity also causes deaths. It's called the energy deathprint, which depends on how the power is generated.

Here are the current choices that our world has:
(arranged from biggest generator of electricity to smallest)

Coal—50% of global electricity
Natural gas
Nuclear
Hydro (dams)
Oil
Wind turbines—1% of global electricity

In order to make a fair comparison of their deathprints, we compare the number of deaths from each source *for the same number of kilowatt-hours produced.*

James Conca's article in FORBES magazine (June 10, 2012) gave this comparison. For every trillion kilowatt-hours of electricity produced, coal causes 170,000 deaths; natural gas causes 4,000 deaths; nuclear causes 90 deaths; hydro causes 1,400 deaths, oil causes 36,000 deaths; and wind causes 150 deaths.

If numbers cause your eyes to glaze over, here's the chart.

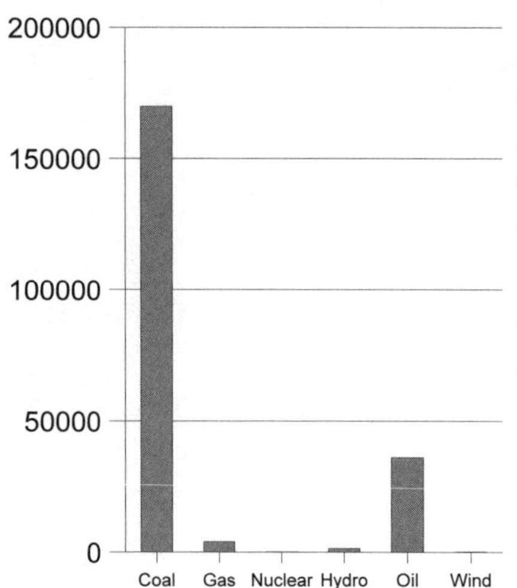

Type `coal deaths vs nuclear deaths` into any computer search. Article after article will say the same thing: electricity generated by nuclear power is by far the safest.

Electricity generated by wind turbines is more than 50% more lethal than nuclear power–and a lot noisier.

end of small essay

Fred continued, "There is one place in chemistry where chemists get to play with protons—a place that is chemistry and not physics. That place is acids.

"Start with a hydrogen atom—one proton and one electron. It's often listed in group 1 (left side) of the periodic table. When it combines with chlorine, bromine, or iodine, we have ionic bonding—the electron is completely stripped from the hydrogen. What's left is H^+—the hydrogen ion, also known as a proton.

"Put some HCl into water and you have a whole bunch of naked protons running around playing in the water. That is the definition of an **acid**.

"Strong acids have lots of naked protons playing in the water. Put a mole* of HCl in water, and you have a mole of protons swimming around.

"Weak acids have few. Carbonic acid, H_2CO_3, the acid in bubbly soft drinks, or acetic acid, CH_3COOH, the acid in vinegar, give up very few of the protons. More than 99% of the H have to stay at home with H_2CO_3 and CH_3COOH rather than go out and play."

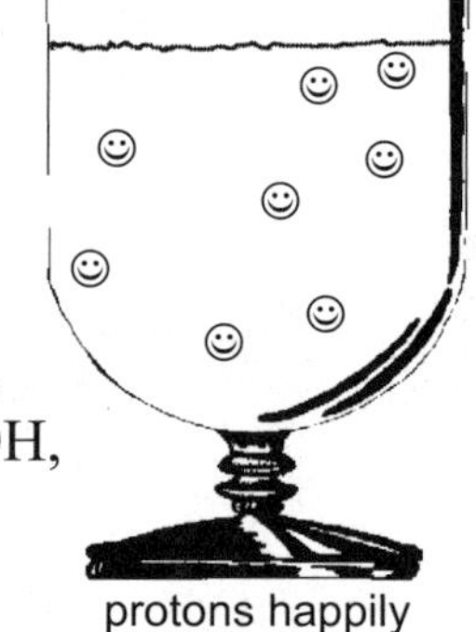

protons happily playing in water

---

* One mole = 602,000,000,000,000,000,000,000 = 6.02×10^{23}

Your Turn to Play

1. We are going to be playing in water for a while. (Fancy chem textbooks call this "investigating chemical reactions in aqueous solutions.")

Stir a handful of salt into a liter of water. This creates a solution. The salt is called the **solute** and the water is called the **solvent**.

Does this give me a liter of solution?

2. Suppose you had a mixture of gold and copper (called red gold in Chapter 12). This is a solution. If it is 18 karat gold (18/24 is gold), then copper is the solute and gold is the solvent. In a solution the smaller quantity is the solute and the larger one is the solvent.

How many karats of gold in red gold would make the gold the solute?

3. Lots of things happen in water. You, my reader, are about two-thirds water.

Chemists need to know how much solute is in a particular solution—how concentrated it is. They want to know how many moles are in each liter of solution. That's called the **molarity** of the solution.

Many chem labs have bottles marked 6 M HCl ("six molar hydrochloric acid").

Atomic weights: H = 1.01 amu
Cl = 35.5 amu

How many grams of HCl are in one liter of 6.00 M HCl?

4. Atomic weight Na = 23.0 amu

The molecular weight of table salt (NaCl) is 58.5.
(The math: 23.0 + 35.5)

If I take 58.5 grams of NaCl and dissolve it in a liter of water, why will this be less than 1 M NaCl?

5. Let's use lots of conversion factors. How many grams of NaCl are in 3.45 liters of 6.00 M HCl to which has been added 7.88 liters of water?

Hint: You can do this problem in your head and give the exact answer without the use of a calculator.

.......COMPLETE SOLUTIONS.......

1. Imagine a full glass of water into which you pour a bunch of salt. The water will overflow the glass if you add enough salt. We have more than a liter of solution.

2. In order for the gold to be the solute it would have to be less than half of the mixture. It would have to be less than 12 karats. Twelve karat gold is exactly half gold (12/24 = 1/2).

3. Molecular weight of HCl = 36.51 (The math: 1.01 + 35.5)

$$\frac{6.00 \text{ M HCl}}{1} \times \frac{1 \text{ mole HCl/L}}{1 \text{ M HCl}} \times \frac{36.51 \text{ g}}{1 \text{ mole HCl}} \doteq 219 \text{ grams/L}$$

This is the definition of molarity—how many moles per liter.

4. One mole of NaCl in one liter of solution is 1 M NaCl.

But we have more than one liter of solution. (See problem 1, above.)

In chemistry, if you want 1 M NaCl, you stick a mole of NaCl (58.5 grams) in a measuring cup and add water until you have a liter of solution.

Actually, in chem labs they don't use measuring cups. Instead they use tall cylinders so that volumes can be measured more accurately than wide measuring cups. Chemists like to use the fancy word *volumetric*.

If you are in the kitchen working with your mother, father, wife, husband, brother, sister, friend, son or daughter, you can ask them, "Please furnish me with a volumetric container." Your mother, father, wife, husband, brother, sister, friend, son or daughter will be impressed with your chemistry language, but they won't know that you are asking for a measuring cup.

5. Reading carefully is really helpful when you sign rental agreements or modeling contracts or loan papers. The rental agreement might require you make your own repairs. The modeling agreement might require that you cut your hair really short. The loan agreement might include some unexpected stuff. I once refinanced my house and actually read the four pages of fine print in the deed of trust. On the third page it stated that everything I owned, not just the house, was the security for the loan. The answer to the question is 0 grams.

Chapter Twenty-eight
Discovering K

Fred held up a plastic water bottle. "Lots of people carry these and like to stay hydrated." Joe made a face and shook his head. For Joe, water was only for rinsing your mouth after brushing your teeth.

"Hydrogen ions also like to stay hydrated. In water, the naked proton is surrounded by polar water molecules. The "O" end is more negatively charged than the "H" ends. You can guess what happens.

H
O
H

H
O—H
H

"All three hydrogen atoms share the oxygen atom equally. In chemistry it is called a **hydrated proton**. H_3O^+ is the **hydronium ion**. When we talk about an acid as lots of naked protons ☺ playing in water, we can be very correct and say *hydrated protons,* H_3O^+, or we can keep things simpler and just say H^+. When you are being interviewed for a job as a chemistry teacher, you should always say hydronium ion. When you're doing chemistry among friends, H^+ is easier to write.

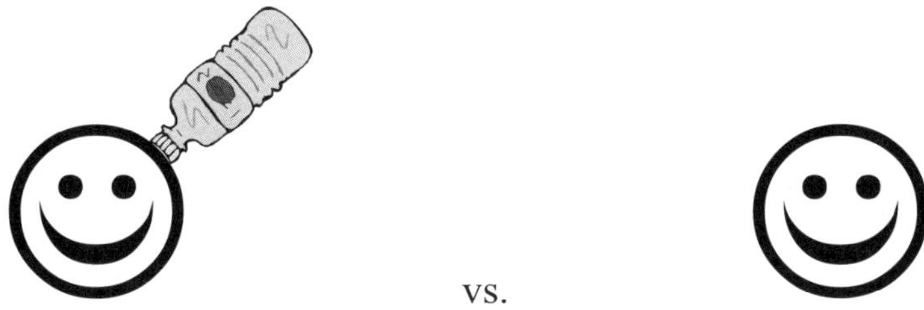

"With strong acids—like HCl, HBr, and HI—the hydrogen atom loses its electron completely—ionic bonding—and runs off by itself to play in the water.

"With weak acids—like carbonic acid (in bubbly soft drinks), acetic acid (in vinegar), or hydrofluoric acid (HF)—very few H^+ ions are in the water."

Two students raised their hands.

Betty asked, "You would think that HF would have ionic bonding like HCl, HBr, and HI. Fluorine, chlorine, bromine, and iodine are all in group 17—the halogens. What makes the difference?"

Fred didn't know. He looked up at the chemistry teacher, Bob Bunsen. Bob said that HF is not a strong acid because it does not completely dissociate in water.

Betty thanked Bob. Fred almost passed out. He thought to himself **The students accepted that answer! I can't believe it. Bob told them that HF is not a strong acid because it does not completely dissociate in water. But that's the definition of a strong acid. In effect, he told them that HF is not a strong acid because it is not a strong acid.** Fred started to breathe heavily.*

Another student asked, "Is there an easy way to tell how many hydrogen ions are running around? Is there a way to tell how strong an acid is?"

This was an easier question. "There are two ways," Fred began. "One way is to add a couple drops of an acid-base indicator that changes color depending on the concentration of hydronium ions present. For example, phenolphthalein changes from colorless to pink when the concentration of H^+ drops below $10^{-9.8}$.

Quick algebra review.

x^{-a} means $\frac{1}{x^a}$ 10^{-6} means $\frac{1}{10^6}$

$10^{5.3}$ is between 10^5 and 10^6.

$x^{a/b}$ means $\sqrt[b]{x^a}$

$10^{5.3} = 10^5 \times 10^{0.3} = 10^5 \times 10^{3/10} = 10^5 \sqrt[10]{10^3}$

"The other way to tell the concentration of H^+ ions is to use a pH meter. It checks how easily the solution conducts electricity, and that tells you how much electrolyte (H^+) is present."

* Later that evening, Fred did some research and found that three HF molecules tended to form H_2F^+ and FHF^- ions that have strong hydrogen bonds. That didn't make a lot of sense to Fred, but at least it was better than "It's not a strong acid because it does not completely dissociate in water."

It was time for Fred to do another chemistry experiment. Because it involved hydrofluoric acid, he was going to do it as a lecture demonstration rather than have the students do it in the chem lab. HF is extremely toxic. It's not because it's a strong acid—it isn't. But HF can pass through the skin and attack the bones in your body.

Experiment #1

Fred dissolved 1 mole of HF in some water and then added water until he had 1 liter of solution. He had 1 M HF (one molar concentration of HF).

Some of the HF dissociates in water. $HF \rightarrow H^+ + F^-$

And some of the $H^+ + F^-$ recombines back into HF.

The equation is written with arrows in both directions.

$$HF \rightleftarrows H^+ + F^-$$

After several moments the solution reached equilibrium—some molecules of HF, some H^+ ions and some F^- ions. This is a dynamic equilibrium. HF is constantly dissociating into H^+ and F^-, and the two ions are constantly combining to make HF.

But when things settled down, Fred measured the acidity of the solution using a pH meter and found that the concentration of H^+ ions was 0.026 M—which means 0.026 moles per liter. The concentration is written as $[H^+]$.

On the board he wrote:

	[HF]	$[H^+]$	$[F^-]$
initially	1	0	0
at equilibrium		0.026	

Looking at $HF \rightleftarrows H^+ + F^-$, we see that if $[H^+]$ gained 0.026, then $[F^-]$ must have also gained 0.026, and [HF] must have lost 0.026.

	[HF]	$[H^+]$	$[F^-]$
initially	1	0	0
at equilibrium	0.974	0.026	0.026

Experiment #2

Fred repeated the experiment, this time starting with [HF] = 0.3 M, $[H^+]$ = 0.3 M, and $[F^-]$ = 0.3 M. He said, "Now everybody will have an equal start."

	[HF]	$[H^+]$	$[F^-]$
initially	0.3	0.3	0.3
at equilibrium		0.02	

Again, Fred measured the acidity and noticed that $[H^+]$ had lost 0.28 M. He then knew what happened to [HF]—it gained 0.28 M—and what happened to $[F^-]$—it lost 0.28 M.

	[HF]	$[H^+]$	$[F^-]$
initially	0.3	0.3	0.3
at equilibrium	0.58	0.02	0.02

Experiment #3

	[HF]	$[H^+]$	$[F^-]$
initially	5	0.2	0.15
at equilibrium	5.11	0.09	0.04

He combined the results of all three experiments:

$$HF \rightleftarrows H^+ + F^-$$

Results of All Three Experiments			
	[HF]	$[H^+]$	$[F^-]$
experiment #1	0.974	0.026	0.026
experiment #2	0.58	0.02	0.02
experiment #3	5.11	0.09	0.04

Fred waited in silence. He wanted to see if any of the students saw the pattern. The students waited in silence, except for Joe who was enjoying a bag of garlic potato chips.

For Fred, numbers were alive. He knew that 18^2 was 324 and that π was approximately 3.1415926535897932384626433832795. When he looked at

	[HF]	$[H^+]$	$[F^-]$
experiment #1	0.974	0.026	0.026
experiment #2	0.58	0.02	0.02
experiment #3	5.11	0.09	0.04

he immediately saw the pattern.

Let's do it slowly in this *Your Turn to Play*.

Your Turn to Play

1. For experiment #1, multiply the concentration of the products H^+ and F^- and divide it by the concentration of the reactant HF.
   In other words, compute $\frac{[H^+][F^-]}{[HF]}$

2. Do the same for experiment #2 and experiment #3.

3. What works for things in water solutions will also work for gases.

$$H_2 + I_2 \rightleftarrows 2HI$$

The two reactants, H_2 and I_2, and the product, HI, are all gases. Here are the results of three experiments.

	$[H_2]$	$[I_2]$	[HI]
#1	0.20 M	0.20 M	1.47 M
#2	0.30 M	0.26 M	2.05 M
#3	0.18 M	0.32 M	1.76 M

Think of $H_2 + I_2 \rightleftarrows 2HI$ as $H_2 + I_2 \rightleftarrows HI + HI$ Compute $\frac{[HI][HI]}{[H_2][I_2]}$

.......COMPLETE SOLUTIONS.......

1. and 2.	[HF]	$[H^+]$	$[F^-]$	$[H^+][F^-]/[HF]$
experiment #1	0.974	0.026	0.026	(0.026)(0.026)/0.974 ≐ 0.00069
experiment #2	0.58	0.02	0.02	(0.02)(0.02)/0.58 ≐ 0.00069
experiment #3	5.11	0.09	0.04	(0.09)(0.04)/5.11 ≐ 0.00070

What Fred had noticed is that no matter what concentrations you started with, the equation $HF \rightleftarrows H^+ + F^-$ had a unique number associated with it. Multiply the concentrations of the products and divide by the concentration of the reactant (when things have reached equilibrium) and you get a constant.

This is called the **equilibrium constant**.

In chemistry, the symbol for the equilibrium constant is K.

3.	$[H_2]$	$[I_2]$	[HI]	$[HI]^2 / [H_2][I_2] = K$
#1	0.20	0.20	1.47	≐ 54.
#2	0.30	0.26	2.05	≐ 54.
#3	0.18	0.32	1.76	≐ 54.

In general, if you have a solution with reactants A, B, and C in equilibrium with products X, Y, Z, and the equation is

$$5A + 2B + 9C \rightleftarrows 3X + Y + 4Z$$

$$\text{then } K = \frac{[X]^3[Y][Z]^4}{[A]^5[B]^2[C]^9}$$

With K = 54 (in problem 3), that means that products in equilibrium are in much higher concentrations than the reactants.

Instead of writing $H_2 + I_2 \rightleftarrows 2HI$, I could be silly and write

$$H_2 + I_2 \rightleftarrows 2HI$$

On second thought, that's really not so silly.

Chapter Twenty-nine

Using a pH Meter

The only thing that Fred had left out in his first experiment was how to read a pH meter. When he tested the acidity of the hydrofluoric acid, he announced that $[H^+]$ was 0.026 M (0.026 moles of H^+ per liter).

Fred wanted to keep it simple when he was showing how K was discovered. With all those numbers flying around

	[HF]	$[H^+]$	$[F^-]$	K = $[H^+][F^-]/[HF]$
experiment #1	0.974	0.026	0.026	0.00069
experiment #2	0.58	0.02	0.02	0.00069
experiment #3	5.11	0.09	0.04	0.00070

he didn't want to add the complication of learning about the pH scale.

But now that the dust had settled about the equilibrium constant K, it was time to show how when the pH meter read `1.585`, Fred could announce that $[H^+]$ was equal to 0.026 M.

Time Out!

Not everyone learns things at the same speed. That is one of the big problems with classroom learning. When I, your author, taught in high school and college, one of the hardest things was providing something for each of the students.

If I just taught at a medium speed, then the bottom one-fifth of the class was lost, and the top one-fifth was bored.

With book learning (the technical name for *reading*), each student can proceed at the rate that feels right. If learning that the product of the products divided by the product of the reactants is a constant, was challenging, then the option of studying that chapter more carefully is available. When I, your author, have been studying math, sometimes I spend the better part of an hour on a single paragraph that originally baffles me. After spending the time, mystery often becomes mastery.

For 0.000032% of my readers—and for Fred—reading a pH of `1.585` and seeing that meant $[H^+]$ was 0.026 M is obvious.

For almost everyone else, a little explanation of pH may help a lot.

For Joe . . . he's happy with the five-pound box of oatmeal cookies that Darlene had baked for him.

That the pH of $[H^+] = 0.026$ is `1.585` doesn't explain pH.

The pH of coffee ($[H^+] = 10^{-5}$) is `5`.
The pH of urine ($[H^+] = 10^{-6}$) is `6`.
The pH of water ($[H^+] = 10^{-7}$) is `7`.
The pH of ammonia ($[H^+] = 10^{-11}$) is `11`.

Can you tell me what the pH of bleach is, if I tell you that the concentration of H^+ ions is 10^{-12} moles per liter?

There are some people's names that I, your author, think are really cool. One that I especially like is Søren Sørensen. He was Danish and that's why his o's looked like this: ø.

In 1909 (five years before the start of WWI) Søren was working as a chemist in a brewery.* He got tired of writing 10^{-4} and of writing 0.0001 all day long. He shortened . . .

10^0	10^{-1}	10^{-2}	10^{-3}	10^{-4}	10^{-5}	10^{-6}	10^{-7}	etc.
to								
0	1	2	3	4	5	6	7	← pH

He got rid of both the 10s and the negative signs. What a relief! The whole world—in science, engineering, and medicine—soon copied him.

* English is crazy. A *brewery* is an establishment that makes beer. I would have called it a beerery—but there is no such word.

One mole of HCl in one liter of solution yields one mole of H^+ ions. $[H^+] = 1$ M (one molar solution). $1 \text{ M} = 10^0$ pH = 0

Very quick algebra review.

$x^0 = 1$ for any x that might come up in a calculation.  $^0 = 1$

So pH = 0 is very acidic.

pH = 14 (translation: $[H^+] = 10^{-14}$) is very **basic**. Basic is the opposite of acidic.

pH = 7, which is the concentration of hydrogen ions in water is considered neutral—neither acidic nor basic.

Joe had tucked a paper napkin in his shirt as a bib and was settling down to a steak dinner. (Joe called it a "steak snack," but that is hard to pronounce. Try: *A steak snack is a snack of steak.*)

This was a perfect opportunity for Fred to talk about how meats get digested.

"What dissolves meat?" Fred asked. "Acid. And stomach acid has a hydrogen ion concentration of 0.0251 M. Meat doesn't have a chance. The pH of gastric juice (stomach acid) is `1.6`."

Two students raised their hands.

The first student asked, "If the stomach acid can dissolve meat, why doesn't it dissolve the stomach wall?"

"The stomach protects itself by producing tons of thick mucus which protects the lining."

Joe wrote in his notes: stomach snot. He wasn't exactly correct, but he was close enough.

Fred said, "When the stomach acid escapes upward into the esophagus, you experience what is called heartburn. When you throw up, part of bad taste is the stomach acid. Your mouth doesn't have the protection that your stomach lining has."

Those words were not the ones that Joe should have heard. He had come to the point where his body was rebelling against the long train of abuses he had inflicted upon it.

He put down his knife and fork. His face was a pale green. He quickly left the room.

"Long train of abuses" is a quote from a very famous document: "But when a long train of abuses and usurpations, pursuing invariably the same Object evinces a design to reduce them under absolute Despotism, it is their right, it is their duty, to throw off such Government. . . ."*

The second student's question was, "I can see how you go from $[H^+] = 10^{-4}$ to a pH of 4. But when you were talking about stomach acid where $[H^+] = 0.0251$, how did you get to 1.6?"

This was a math question, and Fred was in very comfortable territory. When he had been asked why HF was not a strong acid like HCl, he turned to the chemistry teacher, Bob Bunsen, for help.

Now he was asked to find the answer to $0.0251 = 10^?$ and that was child's play for him. If he could show that $0.0251 = 10^{-1.6}$ then using Søren's trick of getting rid of both the 10 and the negative sign would give the pH of 1.6.

Fred wrote on the board in big letters

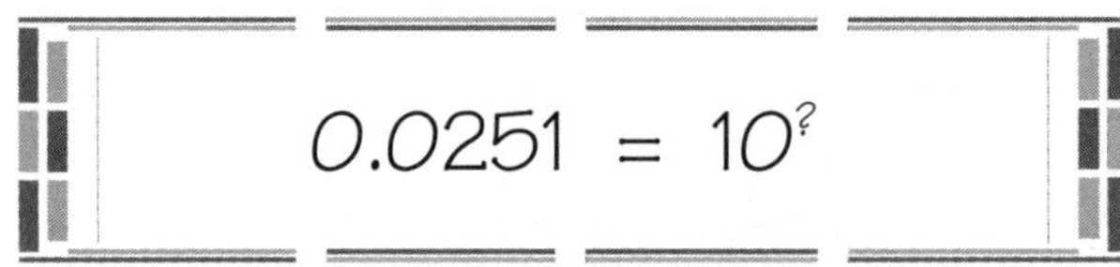

This asks the question, *to what power do you raise 10 in order to get 0.0251?*

You know that $10^{-2} = 0.01$ is too small.

You know that $10^{-1} = 0.1$ is too large.

The answer has to be between –1 and –2.

There are two kinds of calculators. The basic one has only +, –, ×, ÷, and √ keys. Years ago, when they first came out, I, your author, saw

---

* Some students might guess this was from a Confederate document in the Second War of Independence (also known as the Civil War), but it actually came from a document in the First War of Independence, which declared our independence from England.

a full-page magazine ad for one of them. They were about $100. Today, you can pick up a basic calculator for about a dollar when they are on sale.

The second kind of calculator is called a scientific calculator. It has keys marked sin, cos, log, and ln. You'll need one for second year algebra. On those calculators is a 10^x key (located usually right above the log key).

Using the 10^x key, you can try various numbers to find $10^?$ that will equal 0.0251.

Your Turn to Play

Obtain a scientific calculator. We will wait right here until you get one. Just shut this book and, if necessary, go borrow your older sister's. If you are going to have employment that requires more than saying, "Would you like fries or a drink with your order?" you will need a scientific calculator. They are under $20, and I've seen them on sale for under $10.

1. Good. You're back. Let's play with the 10^x key.
   Find $10^{-1.4}$. It will be larger than the 0.0251 we're looking for.
   Find $10^{-1.7}$. It will be smaller than the 0.0251.

2. Play around until you get something close to $10^? = 0.0251$.

3. For $H_2 + I_2 \rightleftarrows 2HI$ the equilibrium constant expression was

$$K = \frac{[HI]^2}{[H_2][I_2]}$$

State the equilibrium constant expression for $CH_4 + 2H_2S \rightleftarrows CS_2 + 4H_2$

4. [harder question] If you started with a 1 L container with $[H_2] = 0.4$ and $[I_2] = 0.2$ and allowed $H_2 + I_2 \rightleftarrows 2HI$ to reach equilibrium, what would be the value of [HI]? You are given that K = 54.

This is labeled a harder question. That means that you will probably not just look at it and write out an answer. Expect to take 5 or 10 or 15 minutes of playing with it. If you just glance at it and say to yourself, "I don't get it" and then turn to the answer, you won't learn as much.

The ability to stick with a problem, rather than taking the easy way out, is a learned ability. The more you practice it, the better you get. Somewhere between the extremes of sloth and agony is the happy middle ground of effort. SLOTH.................effort.................AGONY

.......COMPLETE SOLUTIONS.......

1. $10^{-1.4} \doteq 0.0398$

   $10^{-1.7} \doteq 0.0199$

2. Trying $10^{-1.5} \doteq 0.0316$

   Trying $10^{-1.6} \doteq 0.02511$ which is pretty close to 0.0251

3. The equilibrium constant expression for $CH_4 + 2H_2S \rightleftarrows CS_2 + 4H_2$

   is $K = \dfrac{[CS_2][H_2]^4}{[CH_4][H_2S]^2}$

4. For $H_2 + I_2 \rightleftarrows 2HI$ the equilibrium constant expression is

   $K = \dfrac{[HI]^2}{[H_2][I_2]}$ K = 54 is given

	$[H_2]$	$[I_2]$	[HI]
initially	0.4	0.2	0
change in going to equilibrium	–x	–x	2x
at equilibrium	0.4 – x	0.2 – x	2x

For every mole of H_2 that is "used up" in the reaction,
one mole of I_2 is used up, and
two moles of HI are formed.

Putting 0.4 – x, 0.2 – x, 2x, and K = 54 into the equilibrium constant expression $54 = \dfrac{(2x)^2}{(0.4 - x)(0.2 - x)}$

Multiplying both sides by (0.4 – x)(0.2 – x) $54(0.4 - x)(0.2 - x) = 4x^2$

Multiplying out $54(0.08 - 0.6x + x^2) = 4x^2$

More multiplying $4.32 - 32.4x + 54x^2 = 4x^2$

Put everything on one side and multiply by 100 to get rid of the decimals $5000x^2 - 3240x + 432 = 0$

Quadratic formula $x = \dfrac{3240 \pm \sqrt{(-3240)^2 - 4(5000)(432)}}{10{,}000}$

Doing the arithmetic $x = 0.46$ or 0.1877

x = 0.46 is not possible since that would make the final concentration of $[H_2]$ negative. The final [HI] = 2x = 2(0.1877) = 0.3754

Chapter Thirty
Correcting Fred

Bob Bunsen had something to say. He couldn't stand Fred using just plain old K for an equilibrium constant. He went to the board and wrote K_a next to the HF ⇄ $H^+ + F^-$ reaction because this was an acid ionization equilibrium constant.

He wrote K_p next to $H_2 + I_2$ ⇄ 2HI because this was a reaction involving gases. The "p" was a partial pressures of the gases.

And K_c for equilibriums that are expressed in concentrations.

And K_{sp} for solubility products.

And K_b for base ionization constants.

And K_f for formation constants of complexes.

And K_d for dissociation constants in complexation reactions.

Fred asked, "Aren't all these K's just the concentrations of the products multiplied together and divided by the concentrations of the reactants multiplied together?"

Bob shrugged and said, "Yes, but we gotta keep all those K's separate from each other."

"Why?"

Bob wasn't sure.

Joe had just returned from what some polite people call "bowing to the porcelain god." His tummy felt a little better. He was chewing some bubble gum to get the bad taste of gastric juice out of his mouth. He wrote in his notes:

Bob	Fred
0	1

But Bob wasn't finished. "You forgot to mention **Le Châtelier's principle**. [le-SHOT-lee-ay] Every chemistry teacher who talks about chemical equilibriums* has to mention that principle."

---

* English is tough. There are two permissible ways to pluralize *equilibrium.* The other plural is *equilibria.*

"I didn't forget Le Châtelier's principle. That's just a trivial consequence of equilibrium constant K."

Bob Fred
0 ~~1~~ 2

Fred continued, "If you have a container with hydrogen and iodine gas in it and it has reached an equilibrium $H_2 + I_2 \rightleftarrows 2HI$, everybody knows what would happen if you shove in more H_2.

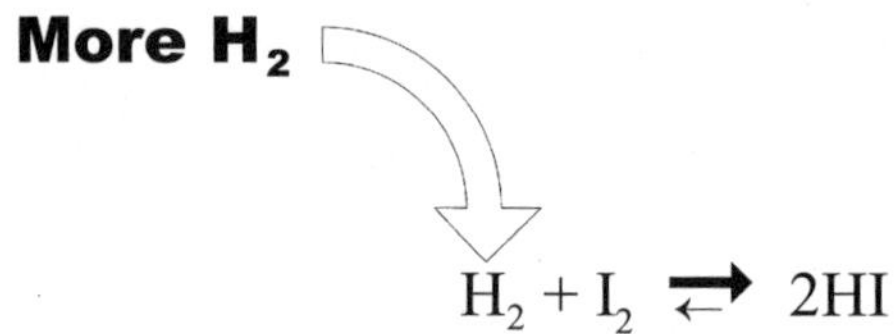

$$H_2 + I_2 \rightleftarrows 2HI$$

"It will drive the equation to the right. If you stuff a whole bunch of people in the front of a bus, some of them will move toward the back of the bus.

"If a lot of people in the back of the bus get off, then people in the front will tend to move toward the back. Le Châtelier's principle is that easy.

"The math is just as easy. We know $K = \dfrac{[HI]^2}{[H_2][I_2]}$ where $K = 54$.

"If I increase $[H_2]$ in the equation $54 = \dfrac{[HI]^2}{[H_2][I_2]}$ that will make the denominator larger. The only way to keep it balanced is to increase the numerator $[HI]^2$ (or decrease $[I_2]$ or both).

Joe wrote in his notes: La Châtelier means if you stick a lot of food on the left side of your mouth and chew, some of it will drift over to the right side.

He tested that by putting a handful of peanuts into his mouth. Then he realized that he still had the gum in his mouth.

small essay

Real Life

Real life is like discovering that you have both gum and peanuts in your mouth.

Translation: Often in real life, you will make mistakes, and that will cut your options way down.

✣ You are at the movies and realize that today is your wedding day and everyone is waiting at the church for you.

How we deal with our reduced options says a lot about who we are.

Do you . . .
- Stay at the movies?
- Rush to the wedding and offer big apologies?
- Leave the country?

end of small essay

Joe had few options. He didn't want to swallow the gum and peanuts. He didn't want to spit into his hand and head back to the restroom to clean up. He borrowed Darlene's handkerchief.

After he did you-know-what into her handkerchief, he tied the corners into a neat bow and placed it on the floor between them.

Now it was Darlene's problem.

Joe had missed a lot of Fred's discussion about bases. Joe had enjoyed learning about acids.

acids

H^+ ions happily playing in water

When Joe was gone, Fred had first introduced **polyatomic ions**, which are ions made up of more than one atom.

OH^-	hydroxide
NO_3^-	nitrate
HCO_3^-	bicarbonate
HSO_4^-	bisulfate
NH_4^+	ammonium

The list could go on for pages. These five are the important ones.

These are like gangs of atoms that like to stick together. The first four (hydroxide, nitrate, bicarbonate, and bisulfate) like to go around stealing an electron. You know who their victims will be—all the group 1 (hydrogen, lithium, sodium, potassium, etc.) and NH_4. After the theft, the victims are H^+, Li^+, Na^+, K^+, and NH_4^+.

Bases are defined as a whole bunch of hydroxide ions running around playing in water.

OH^- ions happily playing in water

"Sodium hydroxide (NaOH) is a strong base. Dissolve a mole of it in water and you get a mole of hydroxide ions (OH^-) swimming around.

"Now suppose you have a clogged bathroom sink. The water won't run down because there's a bunch of hair and gunk blocking the drain. You don't pour a strong acid, like hydrochloric acid (HCl), into the sink because it reacts with many metals."

Joe turned to Darlene and said, "I tried that once. I poured some acid down the drain and it made the drain work just fine. No clog."

Darlene reminded Joe that the reason the water ran "just fine" was because the acid has dissolved his drain pipes under the sink.

Fred continued, "Sodium hydroxide won't hurt the pipes, but it does a nice job dissolving hair. Many of the drain cleaning products are just NaOH, which is sometimes called lye."

Joe thought of an invention that would make him rich. He would call it Joe's Super Dissolver. It would eat into metals. It would dissolve hair and skin and fingernails.

The secret ingredients would be a mixture of hydrochloric acid and sodium hydroxide. Then there would be a ton of H^+ ions and a ton of OH^- ions.

Joe called it his "super acidity-basity solution."

$NaOH \rightarrow Na^+ + OH^-$ completely giving lots of OH^-.

$HCl \rightarrow H^+ + Cl^-$ completely giving lots of H^+.

"What could go wrong?" Joe asked in his "playground voice" that could be heard everywhere in the auditorium. Joe had never learned about whispering.

Fred wrote on the board $H_2O \rightleftarrows H^+ + OH^- \qquad K = \dfrac{[H^+][OH^-]}{1}$

(In writing equilibrium constants, liquid water and solids sitting at the bottom of a solution are set equal to 1.)

"And K for this reaction is equal to $\dfrac{1}{100{,}000{,}000{,}000{,}000}$

"$K = 10^{-14}$ means that the reaction goes strongly to the left. That means that there will be very few hydrogen ions and hydroxide ions playing in the water.

"Mix equal moles of NaOH and HCl and what you get is Na^+ ions and Cl^- ions playing in H_2O. You get plain old salt water.

"Swimming in the ocean won't dissolve you."

Your Turn to Play

1. Fred had written $[H^+][OH^-] = 10^{-14}$ on the board.

In the previous chapter we learned that water was neutral. It was neither acidic or basic. The pH of water is 7.

What is $[H^+]$ in water?

2. What is $[OH^-]$ in water?
3. If we add some HCl to water and the pH becomes 2, what is $[H^+]$?
4. If the pH is 2, what is $[OH^-]$?
5. If we add some NaOH to water and the pH becomes 8, what is $[H^+]$?
6. If the pH is 8, what is $[OH^-]$?
7. Look at the pattern in the previous six questions. Answer these questions almost as fast as you can write:

A) If pH = 3, then $[H^+]$ = __?__ and $[OH^-]$ = __?__.
B) If pH = 4, then $[H^+]$ = __?__ and $[OH^-]$ = __?__.
C) If pH = 11, then $[H^+]$ = __?__ and $[OH^-]$ = __?__.
D) If pH = 0, then $[H^+]$ = __?__ and $[OH^-]$ = __?__.
E) If pH = x, then $[H^+]$ = __?__ and $[OH^-]$ = __?__.

8. Darlene drew a picture of a wedding cake in her class notes. (She did that a lot along with pictures of wedding dresses and bouquets.)

She wrote *Joe* = H^+. She showed that to Joe and asked him to complete *Darlene* = __?__.

Joe pencilled in H^+ and she erased that.

He wrote Cl^- and she erased that.

He wrote a girl and she wrote *a woman* but that wasn't what she was looking for and erased it.

He wrote I give up and she erased that and wrote ____________.

fill in what she wrote

.......COMPLETE SOLUTIONS.......

1. The pH of water is 7 means $[H^+] = 10^{-7}$.

2. Since $[H^+][OH^-] = 10^{-14}$ and $[H^+] = 10^{-7}$, then $[OH^-] = 10^{-7}$.
Note two things: In pure water the amount of hydrogen ions and hydroxide ions are equal, and both are very small.

3. A pH of 2 means $[H^+] = 10^{-2}$.

4. If $[H^+] = 10^{-2}$ and $[H^+][OH^-] = 10^{-14}$, then $[OH^-] = 10^{-12}$.

5. If pH = 8, then $[H^+] = 10^{-8}$.

6. Since $[H^+] = 10^{-8}$ and $[H^+][OH^-] = 10^{-14}$, then $[OH^-] = 10^{-6}$.

7.

A) If pH = 3, then $[H^+]$ = 10^{-3} and $[OH^-]$ = 10^{-11}.
B) If pH = 4, then $[H^+]$ = 10^{-4} and $[OH^-]$ = 10^{-10}.
C) If pH = 11, then $[H^+]$ = 10^{-11} and $[OH^-]$ = 10^{-3}.
D) If pH = 0, then $[H^+]$ = 10^{-0} (which is 1) and $[OH^-]$ = 10^{-14}.
E) If pH = x, then $[H^+]$ = 10^{-x} and $[OH^-]$ = 10^{14-x}.

8. If *Joe* = H^+, then the furthest thing from Darlene's mind would be ionic bonding in which the parties go their separate ways.

She wanted Joe to complete *Darlene* = ___?___ with *Darlene* = OH^-.

With $[H^+][OH^-]$ equal to $\frac{1}{100{,}000{,}000{,}000{,}000}$ that means that very rarely do you find those two wandering off alone.

Chapter Thirty-one
A Second Spearmint

Joe raised his bandaged hand and asked, "When do we ever get to do a spearmint? That's the funnest part of chem is when we get to mix up stuff."

Darlene hit him on the arm. "Hey. We just did an experiment fifteen minutes ago. [Chapter 20] Don't you remember when we mixed salt with silver nitrate and it made that white cloud of silver chloride that was a precipitate?"

Joe was confused. "Did I precipitate in that?"

"No silly. You *participated* in that. You stood right next to me when I mixed the stuff together."*

Fred thought Joe's suggestion was a good one. For some students, mixing together stuff and washing test tubes is the most memorable part of learning chemistry.

"This time," Fred announced, "there will be no eating or drinking in the chem lab. Please leave food and drink here in the auditorium classroom."

Everyone got up and headed down the hallway to the chem lab. Joe left an empty peanut butter jar with a spoon in it.

As they walked, Fred was thinking hard. He needed to think of an experiment that:

1) would be fun

2) would not involve hydrofluoric acid

3) would be something that would be pretty.

Bob could see that Fred was considering what experiment to do and he suggested, "Why not have the students perform a volumetric analysis of an acid of unknown molarity by interacting with a base such as

---

* If you reread Chapter 20, you'll find that Darlene had paired up with Joe, but when it came time to actually *do* the experiment, it seems that she just watched Chris & Pat, Kim & Sidney, Sam & Tracy, Ashley & Kelly, Drew & Morgan, and Betty (working alone). Being next to Joe was too much of a distraction for Darlene. She had forgotten that she hadn't even touched a piece of chem equipment. Joe had forgotten the experiment completely.

sodium hydroxide or potassium hydroxide with a known molarity in the presence of an indicator such as phenolphthalein?" (fee-nol-THAY-leen)

When Bob talked like this to his chemistry students, their ears plugged up, even when they weren't eating peanut butter. Fred's approach was much more compassionate.

When everyone got to the chem lab, Fred began, "You may work alone or in pairs."

It isn't hard to imagine who grabbed Joe's arm. Hint: It wasn't Chris, Pat, Kim, Sidney, Sam, Tracy, Ashley, Kelly, Drew, Morgan, or Betty.

> **Time Out!**
>
> Just for fun, when you, my reader, heard the names of the chem students—Chris, Pat, Kim, Sidney, Sam, Tracy, Ashley, Kelly, Drew, and Morgan—did you picture them as boys or girls?
>
> All those names can attach to either gender.
> Chris—Christine or Christopher
> Pat—Patricia or Patrick
> Sam—Samantha or Samuel

Fred explained the experiment. "At each place is a bottle of 0.200 M sodium hydroxide.

"0.200 M NaOH means 0.2 moles of NaOH in each liter of solution.

"There is also a bottle of hydrochloric acid of unknown molarity. Your challenge is to find how many moles per liter of HCl is in that bottle.

"At each work station are two burets. (burr-ETs)"

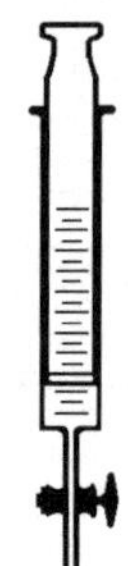

Fred held up one to show how they are used. He explained that you fill one with NaOH and the other with HCl. You don't have to fill them completely—maybe two-thirds full. Then write down how full each one is. For example, the NaOH might be filled to the 50 mL mark. You turn the valve at the bottom to let the base run out. When you are done, you might have a reading of 20 mL. That tells you that you have used 30 mL of NaOH.

Darlene said to herself, "Subtraction." She was right.

"Start by squirting in some HCl, around 30 mL, add a couple drops of phenolphthalein and then add NaOH until the solution turns neutral—the amount of H^+ equals the amount of OH^-."

Betty translated that in her mind, "Till the pH is 7."

Fred finished with, "Then find the molarity of the HCl."

A half dozen hands went up, each with the same question: "How will we know when the solution turns neutral?"

Fred smiled and said, "You'll know."

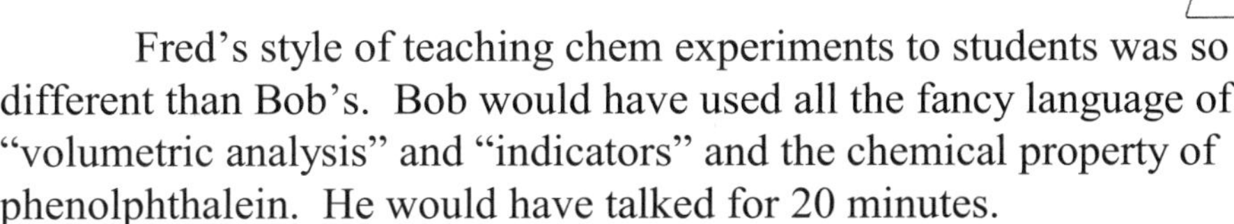

Fred's style of teaching chem experiments to students was so different than Bob's. Bob would have used all the fancy language of "volumetric analysis" and "indicators" and the chemical property of phenolphthalein. He would have talked for 20 minutes.

Fred essentially said, "Squirt and squirt." And because of that, his students were transformed from robots who are instructed step-by-step into thinking human beings.

Historical Note

This experiment is <u>the</u> classical experiment that virtually every beginning chem student does. It's called acid-base titration. (tie-TRAY-shun)

Students learn to be careful and add the NaOH drop-by-drop when they get near the neutral point. They learn to read a buret by looking at the top of the liquid straight on rather than at an angle.

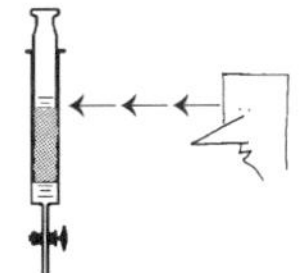

And the students get to work with chemical equations ($HCl + NaOH \rightarrow NaCl + H_2O$), and milliliters and molarity, which is moles per liter of solution.

Darlene squirted some HCl into the flask and added a couple of drops of phenolphthalein. Joe said he wanted to play too. He took the buret with the NaOH in it and shot the whole thing into the flask.

The solution turned pink.

They learned that the phenolphthalein is colorless in acid solutions and pink in basic solutions. If the pH < 7, it's colorless. That's why it's called an indicator.

Darlene squirted some more HCl into the flask until it turned colorless again. Joe refilled the NaOH buret and shot the whole thing into the flask again. It turned pink. He had forgotten to write down how much sodium hydroxide he had used.

Darlene shouted at him, "This isn't a bombing run! The idea isn't just to turn the thing pink."

In contrast, Betty got to the point where one drop of NaOH would turn it pink, and one drop of HCl would turn it back to colorless.

She recorded in her lab notebook: *26.1 mL 0.200 M NaOH balanced 30.0 mL HCl of unknown molarity.*

Ashley got some of the HCl on his* shirt. There would be little holes in his shirt tomorrow. That's why some chemists wear lab coats.

HOW TO DO THE MATH—THREE STEPS

① Find out how many moles of OH^- were used.

Work in liters. $\frac{0.0261 \text{ L}}{1} \times \frac{0.200 \text{ moles NaOH}}{\text{L}} = 0.00522 \text{ moles NaOH}$

② Find out how many moles of HCl were used. 0.00522 moles HCl

Look at the equation. HCl + NaOH → NaCl + H_2O. One molecule of HCl was used for each molecule of NaOH that was used. (Look at the coefficients.)

③ Determine the molarity of the HCl. There was 0.00522 moles of HCl in 0.030 liters. $\frac{0.00522 \text{ moles HCl}}{0.030 \text{ L}} = 0.174 \text{ moles per liter HCl} = 0.174 \text{ M HCl}$

Reading these three steps should not have been at the same rate as when you read about Joe finishing a jar of peanut butter with a spoon.

Study until you get to the point where you can shut the book and recite the ideas behind each step: ① Start with the known buret. Find the number of moles used. ② Use the equation and find the number of moles of the unknown stuff used. ③ Use the definition of molarity.

---

* Did you ever read *Gone With the Wind*?

Your Turn to Play

1. Chris and Pat recorded in their lab notebooks: 36.9 mL of HCl titrated with 32.6 mL of 0.200 M NaOH.

Without looking at the previous page (if you can), do the math and find the molarity of the HCl.

2. If you have 5 moles of H_2SO_4 (sulfuric acid) in 7.5 liters of solution, what is its molarity?

3. How many moles of HNO_3 (nitric acid) are in 9 mL of 6 M HNO_3?

4. Six moles of KOH will neutralize how many moles of H_2SO_4 given the equation $H_2SO_4 + 2\ KOH \rightarrow K_2SO_4 + 2\ H_2O$

5. Fred offered Betty a second titration experiment. He gave her some phosphoric acid (H_3PO_4) and asked her to determine its molarity. She titrated 50.0 mL of it against 16 mL of 0.200 M NaOH.

$H_3PO_4 + 3\ NaOH \rightarrow Na_3PO_4 + 3\ H_2O$

What was the molarity of the phosphoric acid?

H_3PO_4 is called a **triprotic** acid.

tri = three
protic = protons

.......COMPLETE SOLUTIONS.......

1. step ① How many moles of NaOH were used?

$$(0.0326 \text{ L})(0.200 \text{ moles/L}) = 0.00652 \text{ moles of NaOH.}$$

step ② How many moles of HCl were used? 0.00652 moles of HCl.

step ③ The molarity of 0.00652 moles in 36.9 mL is

$$\frac{0.00652 \text{ moles HCl}}{0.0369 \text{ L}} = 0.1767 \text{ M HCl}$$

2. $\frac{5 \text{ moles}}{7.5 \text{ L}} = 0.666 \text{ M}$ This was step ③.

3. $\frac{0.009 \text{ L}}{1} \times \frac{6 \text{ moles}}{\text{L}} = 0.054 \text{ moles}$ This was step ①.

4. Looking at the equation, it takes 2 moles of KOH to balance one mole of H_2SO_4. So 6 moles of KOH will neutralize 3 moles of H_2SO_4.
This was step ②.

5. ① 0.0032 moles of NaOH were used. $\frac{0.016 \text{ L}}{1} \times \frac{0.200 \text{ moles}}{\text{L}}$

② 0.00107 moles of H_3PO_4 were used. $0.0032 \div 3$

③ 0.0214 M H_3PO_4 $\frac{0.00107 \text{ moles}}{0.050 \text{ L}} = 0.0214$

Chapter Thirty-two
I Shall Not Be Moved—Buffers

The class headed back to the auditorium. Joe was the last to get to his seat. The Taffy Truck had pulled up next to the building playing its jingle-jangle tune, and Joe took at side trip to pick a pound of his favorite pink taffy.

When he sat down, Fred was pointing to what he had written on the board before the titration experiment.

$$H_2O \rightleftarrows H^+ + OH^- \qquad K = \frac{[H^+][OH^-]}{1} = 10^{-14}$$

"Suppose $[H^+]$ were equal to 0.00115 M. That's 0.00115 moles per liter of solution. Since $0.00115 = 10^{-2.94}$, the pH is 2.94."

Six hands went up. They all had the same question.* They wanted to know how Fred got from 0.00115 to $10^{-2.94}$ so easily. The only way they knew to do that was to use the 10^x key and trial-and-error.

Try $10^{-2.5}$ on a calculator and get 0.00316.
Try $10^{-2.8}$ on a calculator and get 0.00158.
Try 10^{-3} on a calculator and get 0.00100
Try $10^{-2.9}$ on a calculator and get 0.00126
and so on

Fred let out a secret that only advanced algebra students know. Enter 0.00115 on a calculator and hit the **LOG** key .

It's magic. If you want to know $56 = 10^?$, then enter 56 and hit the log key and get 1.748. So $10^{1.748}$ equals 56.

---

* *Hands had questions?* Once you get beyond See Jane run. See the cat. you learn about figurative language. For adult readers, not all words need to be taken as literally true. It is true that Jane can run fast as the wind, but she can't run 45 miles per hour. It is true that Joe seems as dumb as a box of rocks, but he is really smarter than most rocks.

Six hands went up and they all had the same question is figurative language. This is called metonymy (meh-TAWN-eh-me). Metonymy is naming something by something associated with it. When we read in the newspaper that the White House announced new taxes, this is metonymy. We know that buildings don't talk.

Fred took some 0.00115 M HCl and added a couple drops of phenolphthalein and then added a 0.01 mole of NaOH.

HCl is a strong acid—ionic bonding, complete dissociation in water. Seven moles of HCl will give you seven moles of H^+.

NaOH is a strong base—ionic bonding, complete dissociation in water. Seven moles of NaOH will give you seven moles of OH^-.

"What will happen to the pH = 2.94?"

Joe raised his pink taffy.

He was right. The solution turned pink.

Fred's point was a little deeper* than that. "The pH will plunge from 2.94 down to 12." This was important so Fred wrote it on the board with a box around it.

2.94

↓

12.

drop in pH

Here's the math to show how Fred got the pH = 12. You can skip it unless you consider yourself an A student in chem.

	$[H^+]$	$[OH^-]$
We start with	0.00115	0.01
Change in going to equilibrium	$-x$	$-x$
At equilibrium	$0.00115 - x$	$0.01 - x$

Since $[H^+][OH^-] = 10^{-14}$ (see top of previous page), we have $(0.00115 - x)(0.01 - x) = 10^{-14}$.

That is going to turn into a quadratic equation. That is going to need the quadratic formula. That is something I want to avoid if I can.

A Handy Trick: I know that x can't be bigger than 0.00115, because that would use up all the $[H^+]$. I'm going to **assume** that $x << 0.01$.**

That turns $(0.00115 - x)(0.01 - x) = 10^{-14}$ into $(0.00115 - x)(0.01) \approx 10^{-14}$. A linear equation. Easy to solve. $x = 0.00115 - 10^{-12}$. (I skipped a couple of steps.)

That means, that at equilibrium, $0.00115 - x$ is 10^{-12}. And pH is 12. ☺

---

* "A little deeper" is figurative language—an understatement.

** < means "less than." << means "significantly less than." (Originally, I wrote "a lot less than," but that didn't seem to make much sense.)

"Now, suppose we have a liter of stuff with a pH of 2.94 and we want to add 0.01 moles of NaOH and we don't want the pH to plunge to 12?"

Chris & Pat, Kim & Sidney, Sam & Tracy, Ashley & Kelly, Drew & Morgan raised their hands. They all wanted to know when you would ever *care* about preserving a particular pH. These were ten unusual college students. Most college students never consider the idea of *relevance,* because college teachers often just teach a pile of facts.

Fred smiled and said, "Blood . . . " and then he looked at Joe's hand that had been impaled with broken glass tubing less than an hour ago. His bandage, which the veteran had put on him, was soaked in red. Joe hadn't been very careful.

Fred continued, "Blood has a pH in a very narrow range—from 7.35 to 7.45."

Betty wrote in her notebook, "*Blood is slightly basic.*"

"If you get significantly outside of that 7.35–7.45 range, you will never have to pay taxes again, listen to another commercial on television, pay for another pair of shoes, go hungry, do another homework problem, clean the kitchen, make your bed. . . ."

Joe turned to Darlene and said, "I don't get it."

"What Fred means," she explained, "is that you'll be a dead duck."

"To keep blood in the 7.35–7.45 range, there is a buffer in blood. A buffer keeps a solution in a narrow range."

Joe took notes: Blood buffer makes it tougher. No buffer and I suffer. It's rougher without the buffer.

"To make a buffer, you start with a weak acid. My favorite acid is hydrofluoric acid, HF. You won't find hydrofluoric acid in your blood stream, but HF is a lot easier to write than H_2CO_3, which is in your blood.

"Let's start with 0.25 moles of HF and 0.15 moles of F^- in a liter of water. This is a buffered solution. (You can get 0.15 moles of F^- by putting 0.15 moles of NaF in water.) This gives you a solution with a pH of 2.94." (Fred had carefully chosen the 0.25 moles of HF and 0.15 moles of F^-, so that it would have the same pH as the initial acid solution of HCl on the previous page.)

More stuff that can be skipped (except for A students). Here is how we determined that the buffered solution has a pH of 2.94

	[HF]	[H^+]	[F^-]	
Initially	0.25	0	0.15	
Change in going to equilibrium	−x	x	x	since $HF \rightleftarrows H^+ + F^-$
At equilibrium	0.25 − x	x	0.15 + x	

At equilibrium $K = \dfrac{[H][F^-]}{[HF]} = 0.00069$ **(which was found in the experiment in Chapter 28)**

$$\frac{x(0.15 + x)}{(0.25 - x)} = 0.00069$$

and, again,
assuming $x << 0.15$ $\qquad \dfrac{x(0.15)}{(.25)} \approx 0.00069$

easy algebra $\qquad x = 0.00115$ **(Our assumption of x << 0.15 was right.)**

$x = [H^+] = 0.00115$ implies pH = 2.94 **(using the log key)**

"When we added 0.01 moles of NaOH to the HCl that had an initial pH of 2.94, we had a big drop.

drop in pH

"When we add 0.01 moles of NaOH to the HF & F^- buffered solution that also had an initial pH of 2.94, it made a slight move.

2.94 → 2.99

drop in pH

Here's the vertical strip again to indicate that only A students need to read these computations.

The equilibrium (before adding the NaOH) from the bottom of the previous page was

	[HF]	$[H^+]$	$[F^-]$
equilibrium	0.25– x	x	0.15 + x

and since we found that x = 0.00115 (on the top of this page)

we have	0.24885	0.00115	0.15115 as the starting concentrations.

If we now toss in 0.01 moles of OH^-, that will decrease [HF] by 0.01 and increase $[F^-]$ by 0.01 since the reaction $HF + OH^- \rightarrow H_2O + F^-$ is virtually complete. It is the match up of a weak acid with a strong base.

	[HF]	$[H^+]$	$[F^-]$
Now	0.23885	**(we want to find this value)**	0.16115

From the top of this page $\dfrac{[H^+](0.16115)}{0.23885} = 0.00069$ $\quad$ Therefore $[H^+] \approx 0.00102$

Using the log key, pH = 2.99

"And if I start with my initial buffered solution and toss in 0.01 moles of H^+ (by adding HCl), what's going to happen to the pH?

	[HF]	[H^+]	[F^-]
initially	0.24885	0.00115	0.15115

The extra 0.01 moles of H^+ will almost completely combine with the F^- since $H^+ + F^- \rightarrow HF$ goes almost completely to the right.*

At the new equilibrium	[HF]	[H^+]	[F^-]
	0.25885	x ↙ We want to find this.	0.14115

Your Turn to Play

1. Okay. Do it. Finish it. Find the pH of the resulting solution after you toss in that acid.

I know this question is totally unfair since I told you that you could skip the heavy computations marked with those vertical strips.

But, after all, I've told you that [HF] = 0.25885 and [F^-] = 0.14115 and the formula is sitting in the footnote on this page.

2. Easier question. Looking at the third and fourth lines at the top of this page . . .

	[HF]	[H^+]	[F^-]
initially	0.24885	0.00115	0.15115

is there anything that tells you that hydrofluoric acid is a weak acid?

3. Hydrochloric acid (HCl) is a strong acid. If you have a bottle marked 5 M HCl, fill in the concentrations in this chart:

[HCl]	[H^+]	[Cl^-]
?	?	?

* Since $[H^+][F^-]/[HF] = 0.00069$ means $HF \rightarrow H^+ + F^-$ goes to the left.

.......COMPLETE SOLUTIONS.......

1. Combining the footnote $\frac{[H^+][F^-]}{[HF]} = 0.00069$ with [HF] = 0.25885 and

$[F^-] = 0.14115$ gives the equation $\frac{x(0.14115)}{(0.25885)} = 0.00069$

so $x = [H^+] = 0.001265$. Using the log key, the pH of the solution after adding the acid is 2.898, which is approximately 2.9.

Here are all three boxes.

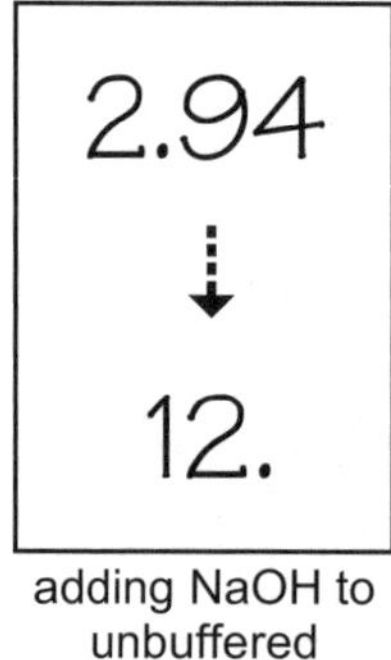

adding NaOH to unbuffered

2.94

2.99

adding NaOH to buffered

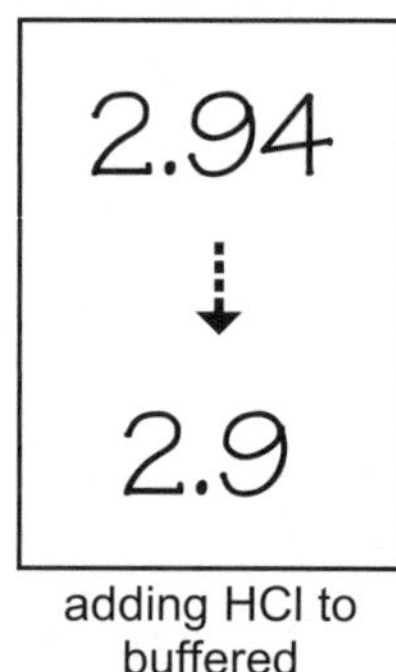

adding HCl to buffered

Buffering works!

2. If HF were a strong acid, a lot more of the HF would dissociate into H^+ and F^-. $[H^+]$ would be larger if HF were a strong acid.

3. HCl does dissociate. $HCl \rightarrow H^+ + Cl^-$

[HCl]	$[H^+]$	$[Cl^-]$
0	5	5

Chapter Thirty-three
Banking on Buffers

Although all the computations with buffers involved only algebra, Fred was concerned that the only message that some of the students got about buffers was like the one Joe wrote in his notes:
Buffers buff.

He needed to draw some pictures so that students could get a feel *why* buffers do their job.

He erased everything on the board except:

Starting with 0.25 moles of HF and 0.15 moles in F^- in a liter of solution, we have at equilibrium

	[HF]	$[H^+]$	$[F^-]$
	0.24885	0.00115	0.15115

He rounded the numbers off so that they would be easier to look at.

we have at equilibrium	[HF]	$[H^+]$	$[F^-]$
	0.25	0.0012	0.15

$$HF \rightleftarrows H^+ + F^-$$

"We have a big 'bank' of HF and a big 'bank' of F^- and a little tiny bit of H^+. It's the H^+ which is critical. It's the pH.

"Suppose we add some more H^+ to this equation. By Le Châtelier's principle that shoves the equation to the left.

more H^+

$$HF \leftarrow H^+ + F^-$$

Some of the deposits in the F^- bank combine with the extra large 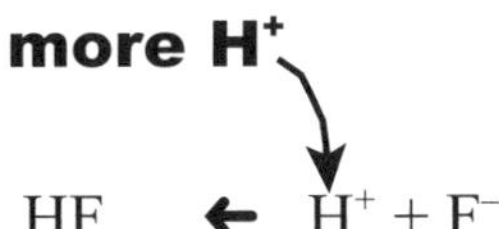 and make some new deposits in the HF bank.

"On the other hand, suppose we add some OH^- to this equation. The OH^- will latch onto some of the (H) ($OH^- + H^+$ make you-know-what).

By Le Châtelier's principle, the equation will be shoved to the right.

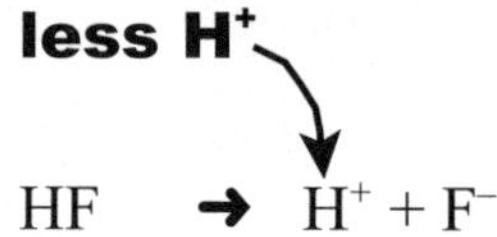

$$HF \rightarrow H^+ + F^-$$

Some of the deposits in the HF bank will be broken apart and given to H^+ and to the F^- bank.

"Le Châtelier's principle with all its pushing and shoving is true because of the equilibrium constant, which we discovered experimentally.

$$\frac{[H^+][F^-]}{[HF]} = 0.00069$$

We can see that the concentrations of HF, H^+, and F^- are locked into each other. I can't change the concentration of one of them without affecting the concentration of the others.

Ashley raised his hand. "Your HF & F^- buffer is only good for a pH of 2.94. What if I want a buffer for, say, pH = 3.25?"

"That's no problem," Fred answered. "My buffer had a pH of 2.94 because I started with 0.25 moles of HF and 0.15 moles of F^-. Just look at that equation again:

$$\frac{[H^+][F^-]}{[HF]} = 0.00069$$

"If you tell me the $[H^+]$ you want and tell me the concentration of HF, that will tell me automatically what concentration of F^- I'll need."

Joe stopped eating pretzels long enough to write in his notes: Fred's Automatic Dial-A-Buffer.

Sam raised her hand. "What if I want a buffer with an initial pH of 2.94—just like the one you started with—but I want it to be able to handle the addition of lots of acid or base. In other words, how do I make a stronger buffer?"

These were perfect questions. Changing the initial pH of a buffer and changing its strength are the two things you can do to play with them.

"I started with 0.25 moles of HF and 0.15 moles of F^- in a liter of solution and the equilibrium equation

$$\frac{[H^+][F^-]}{[HF]} = 0.00069$$

"Suppose I increase the moles of HF to 2.5 and the moles of F^- to 1.5. In other words, suppose I make the concentrations of HF and F^- ten times larger. How does that affect the equation?

"It doesn't. The equation stays in balance.

"But what I've done is to make the two 'banks' ten times larger. They will be able to handle much bigger deposits and withdrawals than before. It's when either bank runs out of 'money' that the buffering stops."

Fred switched from economics back to biology. "When you run (and Joe mentally added to the Taffy Truck) or when you lift weights (and Joe mentally added like a four-pound roast beef sandwich), you breathe hard. How does your body know that your muscles need more oxygen?"

The room was quiet. (Well . . . except for Joe who had been building a stack of pretzels, which just fell down.)

"Less than one percent of adults know the answer to that question. Some parents may tell their kids, 'You breathe harder when you are running because your muscles need more oxygen.' That may be a good thing to tell three-year-olds, but what do you say when they are seven and ask, 'And how do my lungs know that my muscles need more oxygen?'

"It would be really wasteful to put an oxygen sensor on each muscle in the body.

"Instead your circulating blood takes a survey of all your muscles. You only have one pair of lungs. They would get really confused if some muscles were telling them to work harder and others were saying that they could breathe easy.

"The **surprising** thing is that the blood doesn't check how much oxygen the muscles need. All the blood is concerned about is its pH. It's gotta be between 7.35 and 7.45.

"The buffer for blood is carbonic acid and bicarbonate—H_2CO_3 and HCO_3^-. The HCO_3^- takes the place of the F^- that I used when I introduced buffers. The bicarbonate HCO_3^- acts as a single unit.*

"So in the blood, the concentration of H_2CO_3 and HCO_3^- needs to be carefully adjusted to keep the $[H^+]$ in the proper range. It's the old

$$K = \frac{[H^+][HCO_3^-]}{[H_2CO_3]}$$

"We need two more biology facts and everything will become clear.

Fact #1: Cells kick off carbon dioxide (CO_2) when they "burn" food. Carbon dioxide dissolves in water to make carbonic acid.

$$CO_2 + H_2O \rightleftarrows H_2CO_3$$

But when the carbonic acid gets to the lungs, some of the H_2CO_3 breaks down and the carbon dioxide is exhaled. Breathe normally and the concentration of carbonic acid in your blood stays constant.

Fact #2: When you exercise, your muscles pour lactic acid into your blood, and all that extra H^+ puts a lot of stress on K (the equilibrium constant).

"Putting Fact #1 and Fact #2 together: muscles exercising ➠ lactic acid in the blood ➠ $[H_2CO_3]$ is too high ➠ $[H_2CO_3]$ has got to be lowered ➠ exhaling CO_2 and a little help from Le Châtelier makes $CO_2 + H_2O \rightleftarrows H_2CO_3$ move to the left, eliminating some H_2CO_3.

"We breathe hard to get rid of carbon dioxide and restore the pH of the blood . . . and incidentally, in breathing hard we add extra amounts of oxygen into our blood. Just what our muscles needed!

---

* You may remember the list of polyatomic ions including OH^- (hydroxide); NO_3^- (nitrate); HCO_3^- (bicarbonate); and HSO_4^- (bisulfate).

"On the other hand, suppose you get really excited and start breathing really hard. (hyperventilation)

"You lose excessive CO_2. The concentration of carbonic acid in your blood drops. You start to feel lightheaded. The pH in your blood starts to exceed 7.45. It becomes more basic.

"Hey! Did you see that big pizza!"

"Continue heavy breathing and your body takes drastic action. You faint . . . and you breathe slower. Your blood is happier."

Your Turn to Play

1. It's standard first aid procedure to have the victim breathe into a paper bag. Is that procedure used when people have exercised too hard or when they have hyperventilated?

2. The equilibrium constant for $K = \frac{[H^+][HCO_3^-]}{[H_2CO_3]}$ is 4.3×10^{-7}.

Normally, the concentration of bicarbonate is 10 times that of carbonic acid. In that case, what is $[H^+]$ and what is the pH?

3. Oxalic acid is found in spinach. It's also found at higher concentrations in cleaners that will remove rust. Because it is harmful (actually lethal) at high concentrations does not mean it is harmful at low concentrations.

Salt in very high doses is harmful. Without any salt, you die. There is an optimum amount. The same is true for exercise. The same is true for food.

30 mL of oxalic acid ($H_2C_2O_4$) titrates against 42 mL of 0.100 M NaOH.

$$H_2C_2O_4 + 2NaOH \rightarrow Na_2C_2O_4 + 2H_2O$$

What is the molarity of the oxalic acid?

.......COMPLETE SOLUTIONS.......

1. Breathing in a paper bag will cause of buildup of CO_2 in the bag. That will make it harder for H_2CO_3 to turn into CO_2 and H_2O in the lungs.

more CO_2

Le Châtelier at work on $H_2CO_3 \rightleftarrows CO_2 + H_2O$

This will increase the carbonic acid in the blood. That's great when the concentration of carbonic acid has dropped because of hyperventilation.

2. Just algebra.

Given $$\frac{[H^+][HCO_3^-]}{[H_2CO_3]} = 4.3 \times 10^{-7}$$

Given $$\frac{[HCO_3^-]}{[H_2CO_3]} = 10$$

Substitute $$\frac{[H^+]10}{1} = 4.3 \times 10^{-7}$$

Divide both sides by 10 $\quad [H^+] = 4.3 \times 10^{-8}$

Find the pH by using the log key $\quad pH = 7.37$ which is in the range 7.35–7.45

3. There was a three-step approach to titration: ① Find how many moles of OH^- were used; ② Find how many moles of the acid were uses; ③ Find the molarity of the acid.

① Working in liters

$$\frac{0.042\text{ L}}{1} \times \frac{0.100\text{ moles NaOH}}{\text{L}} = 0.0042\text{ moles of NaOH}$$

② 0.0021 moles of oxalic acid were used.

Looking at the equation $H_2C_2O_4 + 2NaOH \rightarrow Na_2C_2O_4 + 2H_2O$, we note that one mole of $H_2C_2O_4$ was used for every two moles of NaOH.

③ $$\frac{0.0021\text{ moles}}{0.030\text{ L}} = 0.07\text{ moles per liter } H_2C_2O_4 = 0.07\text{ M } H_2C_2O_4$$

Chapter Thirty-four
Naming Stuff

Fred looked at the clock. He had been teaching his algebra students and Bob's chemistry students for 50 minutes. He had ten minutes left. At 11 o'clock he was scheduled to teach geometry.

He realized that what he had left out in the first 50 minutes was a lot of names. He was good at explaining *why* things happened. He was good at getting his students to figure out *how* to do things.*

When he was working with gases, it was clear that if you take some gas (under a constant pressure) and heat it up, the volume will go up proportionally. Three liters of gas at 200 K will expand to six liters if you heat it up to 400 K. Knowing that is called **Charles's Law** really doesn't add to a student's understanding. All it does is make the student sound smart at parties.

If you take three liters of gas and double the pressure (at a constant temperature), then it will squeeze it down to 1.5 liters. Pressure and volume are inversely proportional to each other. Everybody knows that squeezing stuff down makes it smaller. Are students smarter if they know that is called **Boyle's Law**?

On the other hand, names are really useful *in situations that are or will be a part of your life.* That woman in your house that has taken care of you since you were born—it's helpful to remember her name. It's Mom.

Learning to drive a car involves a lot of memorizing. The brake pedal is the one on the left assuming there is no clutch pedal. The yellow light means either go ahead or come to a stop depending on a whole host of factors—your speed, how close to the intersection you are, whether there is an idiot tailgating you, whether the roads are dry or icy, etc. But many teenagers memorize all these things because they will be a part of their lives.

---

* Do you remember when he explained titration as "squirt and squirt"?

How much of the culinary arts do you learn? Are you able to take a can of chili, open it, put it into a container, and microwave it? Some people can even put it into a pot and heat it on the stove, but for many, that would be Advanced Cooking—something that they would never need to learn.

What about chemistry? How much of it do you *need*?

You will remember
what you think you will need.

It's that simple.

But knowing *what* you will need is very hard in today's world. If you have gotten this far in your education, there is a good chance that you will be in three-to-five different fields/occupations/endeavors in your life—each one of which will require significant new learning.

I, your author, who has lived most of his life in a world much less turbulent than the one you are entering, have been a . . .

high school and college teacher
author of books on math, biology, physics, economics, prayer, chemistry, language arts
rabbit "rancher"—120 rabbits in the 1980s
investor in naked puts and calls
owner of apartment houses (and doing some plumbing, electrical, carpentry)
publisher
real estate broker
sewer of skirts and blouses for my daughter
Sunday morning guest preacher in more than a dozen different churches
chicken "rancher"—150 chickens in the 2000s.

I took high school and college chemistry courses in the 1960s. I learned that the alkali metals (lithium, sodium, potassium, . . .) are located in the left column of the periodic table and all have a single valance electron that they easily give up. I could never have guessed that fifty years later I would need that as I write *Life of Fred: Chemistry.*

I remembered.

Recently, I looked back on my college transcript and discovered that I had taken a course entitled History of Journalism. That frightened me a little. I can't remember having taken that course. Nothing. It didn't stick. That might have come in useful. I forgot to include . . .

third highest-paid on the staff of the ***Daily Californian*** newspaper.

How much chemistry will you need to remember? Will any of the three-to-five different fields/occupations/endeavors require that you know what a mole is? (Will you have kids who need help with their chemistry?) These are questions that are almost impossible to answer. I'm sorry. I didn't invent modern life. If you lived 300 years ago, the question of what you need to remember would be easy: You would be doing the same thing that your father or mother had been doing.

With that prologue and ten minutes left for Fred to teach chemistry, we face organic chemistry.

Fred was thinking of starting by asking, "What's the difference between a rock and an orange?" but he was afraid Joe would answer, "Oranges taste better."

Fred wanted to highlight the difference between inorganic chemistry and organic chemistry.

For chemists, organic chemistry deals with compounds containing carbon.* **

"When we looked at the alkali metals (Li, Na, K, . . .), which each had a single valance electron that they were happy to give up, and paired them up with the halogens (F, Cl, Br, I, . . .), which each loved to get one more electron, things were simple.

"It was like waltzing—one alkali with one halogen: LiF, LiCl, NaCl, NaI, KF, KBr, and so on."

It was like Joe-Darlene Darlene thought. **Obvious, simple, and perfect.**

"Things got a little more exciting with the group 2 (second column) elements of the periodic table.

1	2	3	4	5	6	7	8	9	10	11	12	13	14	15	16	17	18
H																	He
Li	Be											B	C	N	O	F	Ne

---

* Actually, it is almost always carbon and hydrogen. Carbon dioxide isn't usually included in organic chemistry.

** (A double footnote—is this legal?) When we get to looking at how organic chemistry works inside of living things (DNA and carbohydrates), then organic chemistry turns into biochemistry.

Na	Mg											Al	Si	P	S	Cl	Ar
K	Ca	Sc	Ti	V	Cr	Mn	Fe	Co	Ni	Cu	Zn	Ga	Ge	As	Se	Br	Kr
Rb	Sr	Y	Zr	Nb	Mo	Tc	Ru	Rh	Pd	Ag	Cd	In	Sn	Sb	Te	I	Xe
Cs	Ba														Po	At	Rn
Fr	Ra																

"Beryllium, magnesium, and calcium were all happy to give up two valance electrons. Any of these could dance with two halogen atoms: MgF_2, $CaCl_2$ and so on."

Darlene thought I hope Joe isn't listening to this. He wasn't. He was busy making a smoothie in a blender (6 jelly beans, 2 marshmallows, 1 cup Sluice). Fred had to stop speaking when Joe pushed the liquify button.

Wait! I, your reader, have caught a mistake. Back in Chapter 27, you, Mr. Author, wrote that Joe had 14 Sluice bottles and they were all empty. How did he suddenly get a cup of Sluice?

The big yellow extension cord that Joe ran across the classroom was hooked up to two things: his blender and the Sluice machine, which contained five gallons of fresh Sluice. Joe ordered that machine so that he wouldn't run out again. It was installed right next to his popcorn machine mentioned in Chapter 15.

Thank you. I feel better now.

Fred continued as Joe stuck a straw into his smoothie. "On the other hand, things get really complicated when we get to one particular element in group 14. Carbon. If you make a list of all the known compounds, more than 90% of them will contain carbon. It is the superstar.

"We never wrote Na—Cl because there never was a polar covalent bond between them. Na^+ and Cl^- both ran off and played by themselves. This was ionic bonding."

Darlene mentally scratched off the idea of the Joe-Darlene bond being like sodium chloride.

"In contrast," Fred said, "the C—C bond is very strong, and you can make very long molecules. Earlier this hour I mentioned all the "anes"—methane, ethane, propane, butane, pentane."

"Here is the structural formula for pentane.

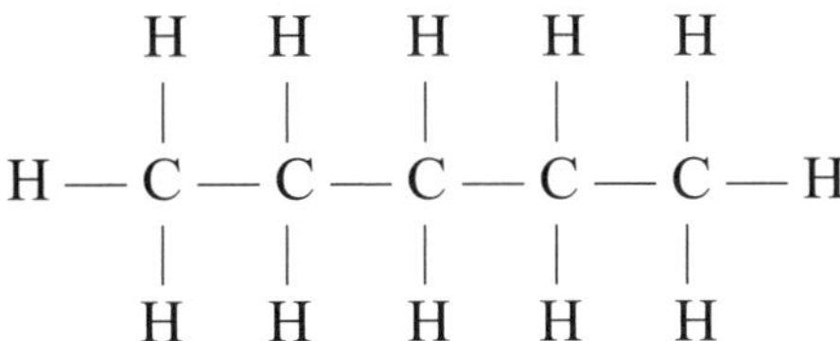

The *pent* means 5 carbon atoms in a row. The *ane* means that there are only single bonds between the C's.

"The nice thing is that the prefixes are used over and over again in organic chem. I've already mentioned *meth, eth, prop, but,* and *pent* for 1, 2, 3, 4, and 5.

"*Hex, hep, oct, non,* and *dec* are 6, 7, 8, 9, and 10."

A big slurping sound echoed in the room as Joe used his straw on the last bit of smoothie. (← That's impolite.) He raised his straw and asked, "What's 20?" Darlene hit him on the arm—both for the silly question and for getting drops of smoothie in her hair.

"In case you ever need it," Fred answered, "it's *eico.*"

Joe opened up a large bag of Blackmallows—licorice-flavored marshmallows. ● He stabbed one of them with a toothpick and Fred shouted, "Perfect! Please bring them up here."

Joe couldn't figure that out. He had never seen Fred eat anything. Fred cleared off a big space on the lab table.

Fred asked, "Do you have any other flavors?"

"Yes. I've got some Dustmallows. ● They are smaller, but they don't taste as good. You can have them if you want."

Fred built an ethane.

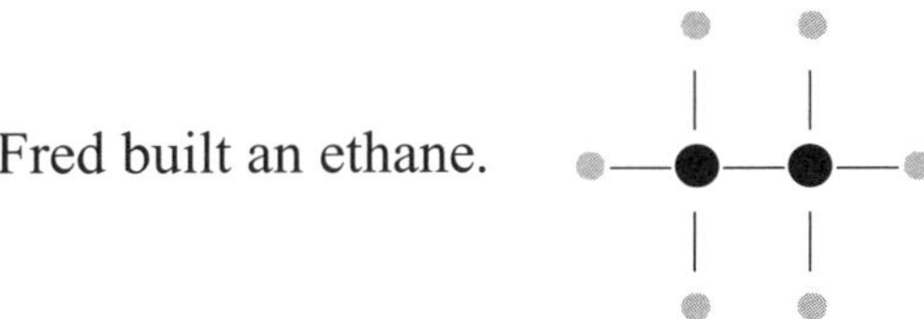

Four toothpicks have to come out of every Blackmallow (carbon).
One toothpick has to come out of every Dustmallow (hydrogen).

"This is the chemistry of hydrocarbons." Joe got really excited.

Fred stood back. Joe wanted to play.

Joe made what he called a caterpillar. C_9H_{20}

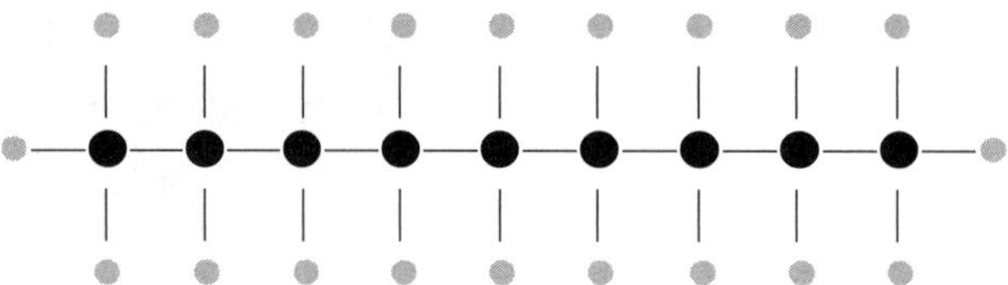

The class shouted, "Nonane!"

Joe got bored with regular caterpillars. He gave it a left turn.

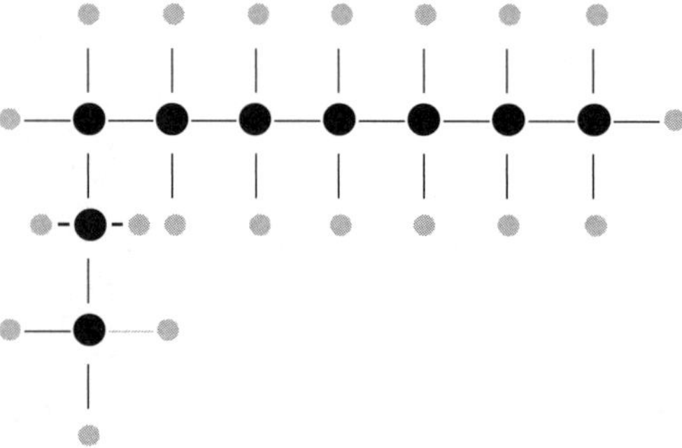

The class was quiet. Fred explained that giving a caterpillar a twist around a single bond doesn't affect anything. It's the same compound.

The class shouted, "Still nonane!"

Joe was having fun. He rearranged his Blackmallows and Dustmallows.

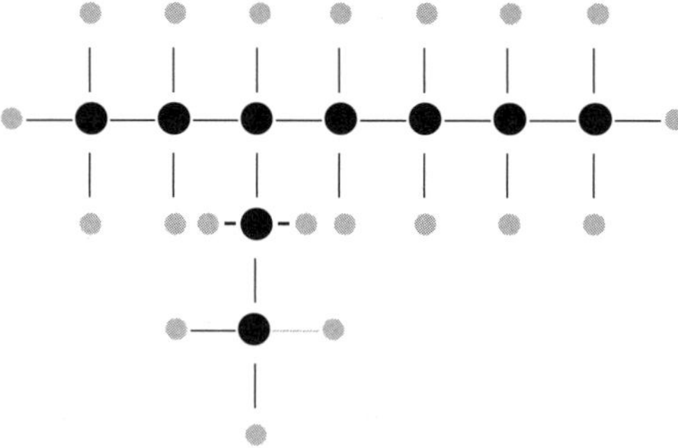

This was no longer nonane. This was a different compound with different physical properties. For example, its melting point was different.

This was no longer a caterpillar (or what chemists call a **straight-chain hydrocarbon**).

The straight part has seven carbon atoms—a heptane. One of the hydrogen atoms (on the third C of the heptane) was replaced by an ethane.

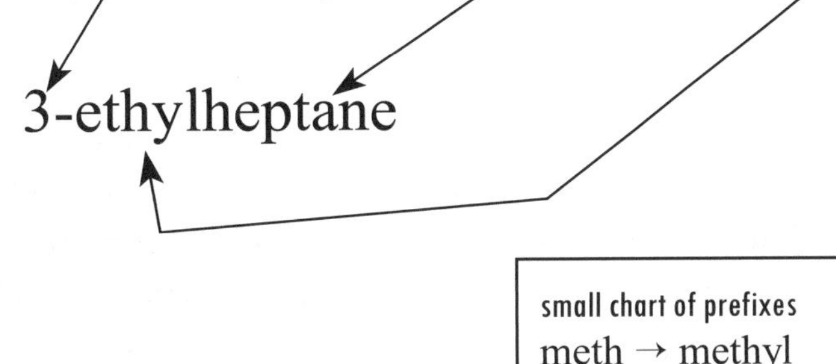

Ethane becomes *ethyl* as a prefix

small chart of prefixes
meth → methyl
eth → ethyl
prop → propyl
but → butyl
pent → pentyl

Look at this until you spot the pattern.

Joe had hundreds of bags of Blackmallows and Dustmallows so everyone in the class could play.

Fred said, "Give me 2,4-dimethylhexane."

Many of the students correctly started by laying out a straight line of six C's. ●—●—●—●—●—● ← hexane

di means *two* in lots of English words. For example, *diadromous* fish are fish that migrate between fresh and salt water. Or *diastema* = the gap between two adjacent teeth. Everyone knows that if you have significant diastemata (diastemata = the plural of diastema), you have an easy time flossing.

Stick a methyl at the second and fourth C in the hexane,

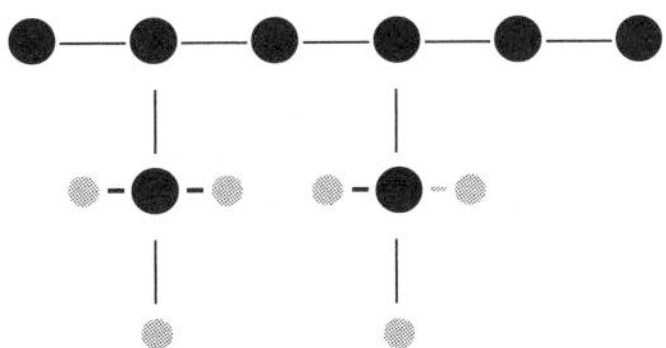

and then fill in all the —● for the ●s that don't have four sticks.

It's is okay if your diagram looked like this:

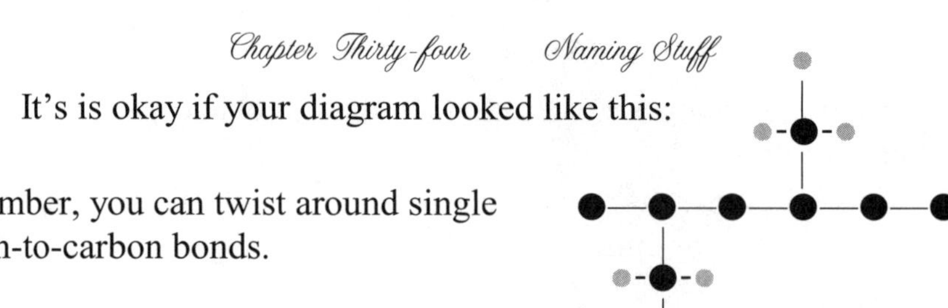

Remember, you can twist around single carbon-to-carbon bonds.

So far, most of the students were following the discussion. (Joe had lit the Bunsen burner and was making s'mores with the Blackmallows and some chocolate he always carried in his pocket.)

To show his students how complex things could get, Fred asked them to start with 5-butyldectane. That was easy. They laid out a string of ten carbon atoms and put a butyl —●—●—●—●—● on the fifth C.

The butyl is substituted for one of the H's on the fifth C. It is called a **substituent group**.

Q: What if the substituent group was ?

A: Then it would be called *sec*-butyl.

Q: What if it was ?

A: It would be called isobutyl.

Q: What if the substituent group was ?

A: It would be called *tert*-butyl.

There are four butyl substituent groups.

Fred's last question: "What's the stick diagram for 6,7-diisobutyl-3-methyldecane?"

(If you have more than one substituent group, they are alphabetized.)

Chapter Thirty-five
More Naming

Things were just starting to get complicated. Fred held up a hydrocarbon and asked the class to name it.

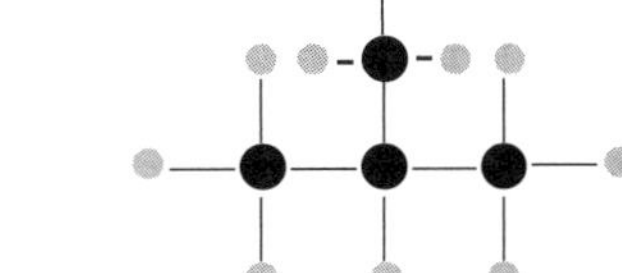

The straight chain was propane.
The substituent group was methyl.
It was attached to the second C in the propane.

The class shouted, "2-methylpropane!"

Fred shook his head. "Chemists call it methylpropane."

Ashley raised his hand and asked, "But how do they know the methyl is attached to the second C unless you say it?"

Fred rearranged the Blackmallows and the Dustmallows.

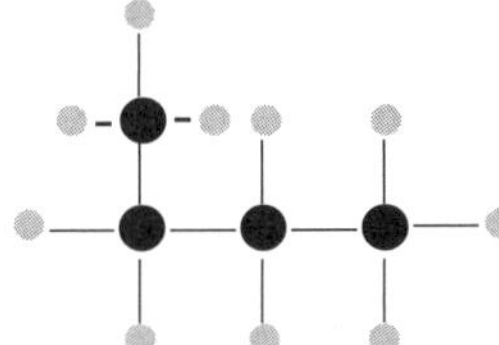

Fred said, "You can't have 1-methylpropane. Remember you can twist around any single bonds. This is just butane—four carbons in a row."

Meanwhile Joe was busy playing with his ●s and ●s on the lab table. The rules were simple:

Four toothpicks have to come out of every Blackmallow (carbon).

One toothpick has to come out of every Dustmallow (hydrogen).

Joe created . . . ●=●—●—●—●

and then he filled in the hydrogens in order to follow the rules . . .

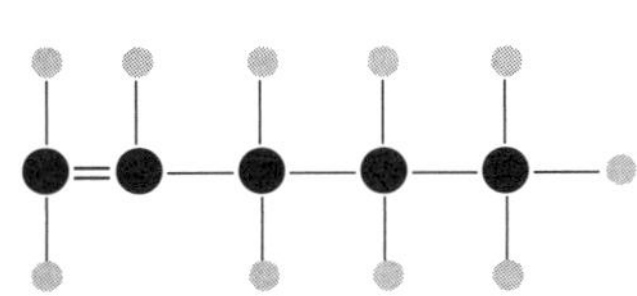

Instead of pent*ane*, this is called pent*ene*. This is 1-pentene.

2-pentene looks like . . .

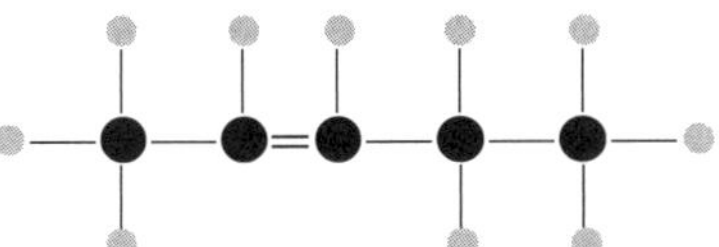

All of the "anes" (such as methane and ethane) are called **alkanes**. If there is a double bond, they are called **alkenes**. If there is a triple bond, they are called **alkynes**.

This is propyne . . .

Happily, to the joy of chem students over the years, alkane, alkene, and alkyne are in alphabetical order.
① ② ③

Recall, we used *meth, eth, prop, but, pent, hex, hep, oct, non,* and *dec* as prefixes in front of *ane, ene,* and *yne* to indicate how many C's were in the main spine of the caterpillar.

Three pages ago we introduced *di* as the prefix for **substituents**.* The *di* in 2,4-dimethylhexane indicated we had two methyls. Your logical question at this point is, "What are the rest of the prefixes? Suppose I have three or four methyls?"

Here's the official list: *di, tri, tetra, penta, hexa* for 2, 3, 4, 5, 6. You can also use these prefixes to indicate the number of multiple bonds. Here is a rule when you have alkenes or alkynes.

If you have double or triple bonds (–enes or –ynes), start your numbering of the spine from the end that is closest to the first multiple bond.

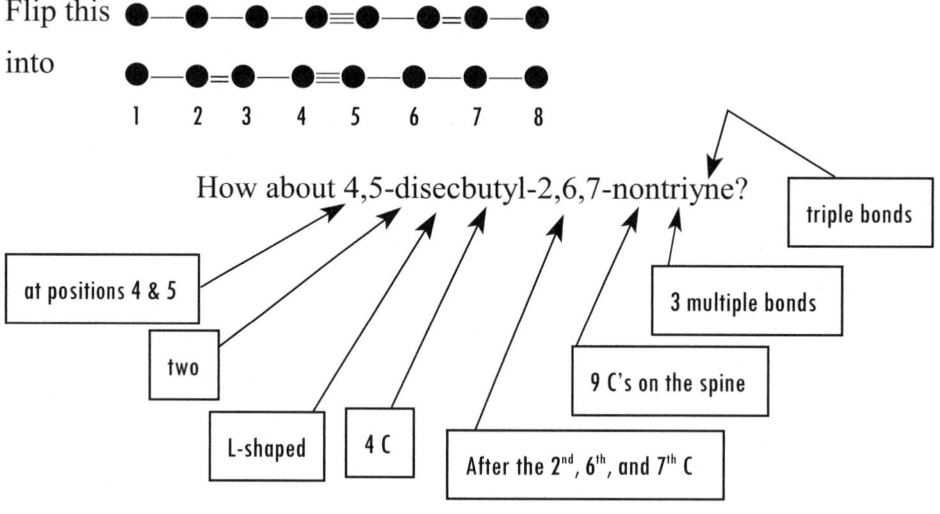

---

* Substituents are the different substituent groups. (It's nice to have two different ways to refer to —●—●, which is ethyl.)

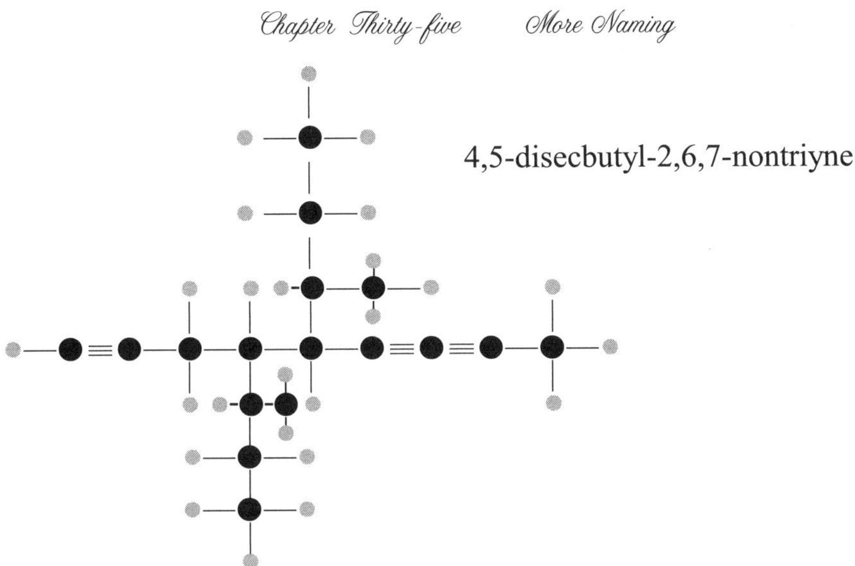

Author's note: It took me 35 minutes to create this drawing. I only made two giant mistakes in those 35 minutes. (Hopefully, there is not a third one that I missed!) If you would spend a minute or two comparing the name with the drawing, we will be able to do some serious talking.

Some Serious Talking between Author and Reader

It's time that you and I had a heart-to-heart talk about this naming stuff in organic chem.

Hey! I was just about to break in after I looked at that "thing" at the top of this page.

Oh? What did you have in mind?

I wanted to throw up. This just seems like memorizing nonsense. I'll never remember it. I figure I'll never use it. Tomorrow I won't be able to tell a butane from a butyl.

First of all, I haven't memorized a lot of this stuff. I do remember alkane, alkene, and alkyne are are single, double, and triple bonds because of the memory aid trick.

What trick?

They are in alphabetical order.

But let's get back to our serious talking.

Okay. Talk.

I don't know how to express this. Let me tell a story.

You are good at stories.

Many years ago I had a girlfriend who was nuts about plants.

A botanist.

Yes. And we went on a walk around the six acres I owned outside of Santa Rosa. She pointed out some plant and told me its name. And a second plant and its name. And a third. . . .

By the end of the hike, she had named at least 20 of them. I remembered her name but none of those plant names. And those plant names would probably be ten times more useful than 4,5-disecbutyl-2,6,7-nontriyne.

Are you trying to tell me that you are chickening out on this organic chemistry thing?

I'm really tempted.

Speak up! Are you a man or a mouse?

Well, why do you have to do this naming of organic chem stuff? And, more particularly, will I have to memorize it?

Did you notice that I didn't put a *Your Turn to Play* at the end of the previous chapter? At this point, turn off your study mode and just enjoy the rest of the organic chemistry. I am required by tradition to stick carbon-based chem in a beginning chem course.

And does that mean . . . does that mean . . . I can't believe what you are telling me.

If I tell you to skip it, I'll get in trouble. If you skip it, you'll get in trouble. How about reading it—every word of it—but *lightly*?

How much is left?

About a ton. Joe will really mess up the Blackmallows and turn from aliphatic hydrocarbons to aromatic ones (called **arenes**). Later we got to pop off the H's and substitute some O's, N's and halogens creating nine classes of compounds.

I have an idea. You have blown about two pages in this chapter with our serious talk. How about just plowing through the stuff and we get all the organic chem finished in this chapter and no Your Turn to Play?

As you wish. (← famous line from the movie "The Princess Bride")

Joe took a string of six Blackmallows and shoved the ends together. He was thinking of a doughnut. Darlene, of course, saw a wedding ring. He stuck some more toothpicks in to make three double bonds.

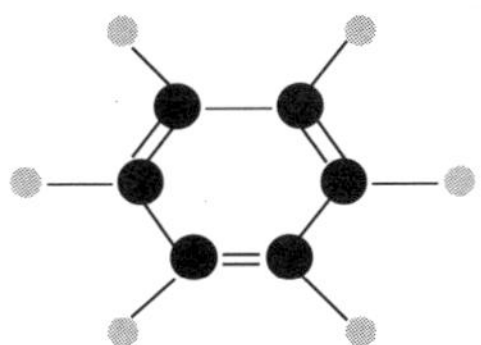

Names: This is called **benzene** or **cyclohexane**. When the spine was straight, those were **aliphatic hydrocarbons**. In a circle, they are the **arenes** or the **aromatic compounds** or **ring compounds**.

Benzene is often written as leaving off the s and representing all the single and double bonds by a circle. (In actuality, there aren't single and double bonds between the six carbon atoms. The six all share everything equally. The circle inside the hexagon reminds us of this. This makes the arenes especially stable.

If we substitute a methyl for one of the hydrogens in benzene, we get CH_3

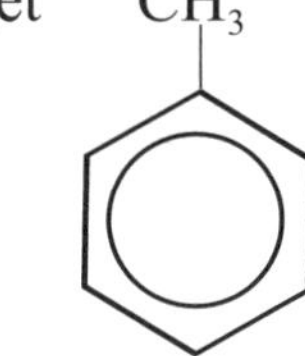

It is called methylbenzene.

There are three different ways to stick two methyls on benzene.

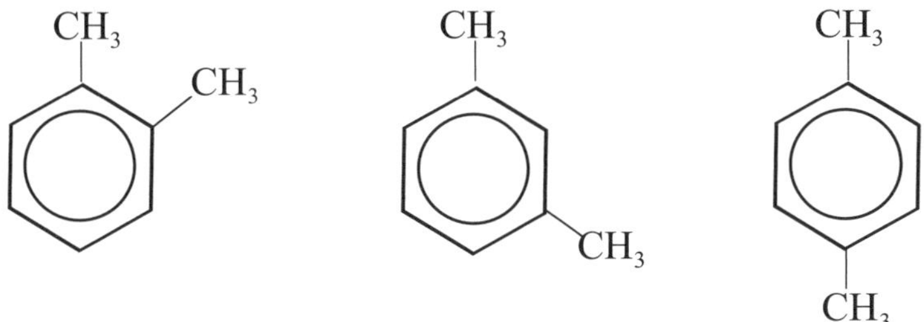

1,2-dimethylbenzene 1,3-dimethylbenzene 1,4-dimethylbenzene

I, your reader, have to interject. I know this chapter is going to be long because of our agreement, but I have to admit that 1,3-dimethylbenzene isn't as frightening as it might have been ten minutes ago.

I didn't get a chance to mention that since chemists so often have substituents on a benzene ring, that they often abbreviate substitutions at carbon atoms 2, 3, and 4 by the prefixes ortho-, meta-, and para-.

So your favorite 1,3-dimethylbenzene can be abbreviated as meta-dimethylbenzene.

Wait a minute! The abbreviation is longer than the original!

Oh well. That's chemistry.

Because those three dimethylbenzenes are so frequently encountered, they are also called **xylenes**, which is great if you are playing Scrabble.® Your favorite 1,3-dimethylbenzene can really be abbreviated as *m*-xylene.

If I just briefly mention naphthalene and anthracene

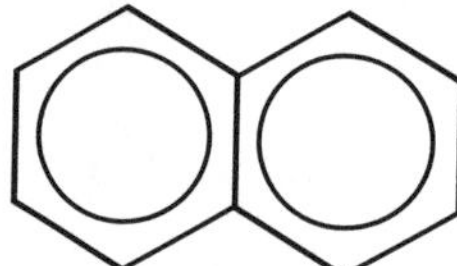

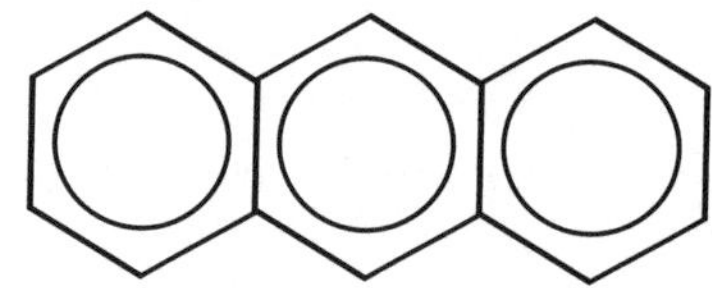

we are now done with hydrocarbons.

We're done?

With hydrocarbons. Now we get to stir in some O's, N's and halogens and create nine classes of organic compounds.

Nine!

Nine.

Classes? That sounds much different than nine compounds.

Yup. When World War I began in 1914, many believed that the troops would "be home by Christmas." That turned out to be true, but it was Christmas of 1918. (The end of the Great War, as it was called back then, was at 11 o'clock on the 11th day of the 11th month.)

I can assure you that we will be done with organic chemistry and all of its naming "by Christmas."

To make things go a little quicker, we are going to get rid of the Blackmallow carbons ● and the Dustmallow hydrogens ●. From this point onward it's C's and H's.

Here are all the rules that we'll need to finish out the chapter:

Older rules

Four toothpicks have to come out of every Blackmallow (carbon).

One toothpick has to come out of every Dustmallow (hydrogen).

If you have double or triple bonds (–enes or –ynes), start your numbering of the spine from the end that is closest to the first multiple bond.

New rules

One toothpick for halogens (F, Cl, Br, and I). —Cl

Two toothpicks for oxygen. —O— or =O

Three toothpicks for nitrogen. —N— (with a third bond downward)

Now it is just a matter of toothpicking together the hydrocarbons with Cl's, O's, and N's and see what we get.

Class #1: The Haloalkanes

This is the easiest. Take an alkane (single-bond hydrocarbon) and replace one or more of the H's with F (or Cl or Br or I). If the label reads 1,2-dichloro-1fluoroethane, you could draw the diagram. Start with the ethane (C–C) and do the decorating (like putting balls on a Christmas tree).

```
F–C–C–Cl
  |
  Cl
```

and stick H's everywhere needed so each C has four toothpicks.

Class #2: The Amines

Nitrogen with C's and H's attached. Now we get some interesting stuff.

NH_3 is ammonia, a gas. In water it becomes a super cleaner for around the house. You can buy it in the grocery store under a zillion different brand names. It's a window cleaner, a kitchen counter cleaner, an electric switch plate cleaner. It will cut into grease, hair spray, and floor wax. You can buy it in bottles that are simply labeled as household ammonia. Take a little wiff of some household ammonia and you'll remember the smell for the rest of your life.

CH_3NH_2, which you could have guessed is methylamine, is the smell of "ripe" fish. If ammonia is stink-level 1 (I just made that up), then methylamine is stink-level 2.

$NH_2(CH_2)_5NH_2$ is the shorthand for N–C–C–C–C–C–N (and stick H's everywhere so C has four toothpicks and N has three). This stuff is stink-level 10. Small story: My dad used to own some apartment houses in San Francisco. Once at the dinner table, when I was a kid, he said that apartment 8 was now empty. Some other tenants had reported to him that the hallway near that apartment really smelled bad. No one has seen the occupant of apartment 8 in weeks. My dad investigated. He knocked, and when no one answered, he entered. The smell of N–C–C–C–C–C–N was overpowering. N–C–C–C–C–C–N has a chemical name, but its common name is much more descriptive: cadaverine. We then went back to eating dinner.

Class #3: The Alcohols

This can be a sensitive subject.

There are lots of people on each side of the issue of grain alcohol—beer (3–5%), wine (10–13%), distilled liquors (35–90%). The Federal government makes a lot of money taxing alcohol. On the other hand, in my family clan in the generation before me, there were two alcoholics.

Noah was a drunk. Years after his ark ride he drank so much wine that he passed out naked in his tent. (Gen 9:21) This isn't mentioned very often in church.

The Centers for Disease Control and Prevention (known as the CDC) has a fact sheet, which can be easily found typing in **alcohol related deaths** in any computer search engine. It reports that misuse of alcohol is the third leading lifestyle-related cause of death in the U.S.

Misuse is associated with . . .

* two-thirds of intimate partner violence
* miscarriage, stillbirths, and mental birth defects that last a lifetime—*any* drinking by pregnant women is considered excessive
* stroke, heart attacks
* depression and suicide
* cancers of the mouth, throat, colon, and breast
* traffic injuries, falls, and burns.

There is a lot of pressure in our society to have a drink. Each individual gets to decide whether he or she will drink. (I, your author, decided never to start. I saw too many negatives.)

To make an alcohol, you take a hydrocarbon and replace one of the H's with an OH.

If you start with methane C, you get methanol C–O–H.

If you start with ethane C–C, you get ethanol C–C–OH.

If you start with heptane C–C–C–C–C–C–C, you get heptanol C–C–C–C–C–C–C–O–H.

The "anes" become "anols."

Ethanol is the grain (drinking) alcohol.

Methanol is really not for drinking. Low doses cause blindness.

Class #4: The Aldehydes

Start with a hydrocarbon and replace one of the H's *at the end of a carbon chain* with a double-bonded O.

If you start with butane C–C–C–C, you get butanal C–C–C–C=O. (As usual, you stick H's on the C's so that each C has four toothpicks.)

"-ane" become "-anal."

The simpliest aldehyde, methanal, C=O, is used in biology labs as a preservative. Biologists shorten the name methanal to formaldehyde.

Class #5: The Ketones

Start with a hydrocarbon and replace one of the H's *in the middle of a carbon chain* with a double-bonded O.

If you start with propane C–C–C, you get propanone C–C–C (with O double-bonded to the middle C)

"-ane" becomes "-one."

Propanone (also known as acetone) is used as a nail-polish remover.

The C=O group appears in aldehydes, ketones (and three other classes). Chemists call it the **carbonyl group**. (car-bo-NEEL)

Class #6: The Carboxylic acids

Start with a hydrocarbon and replace three of the H's *at the end of a carbon chain* with a double-bonded O and a single-bond O.

If you start with ethane C–C, you get ethanoic acid

```
C–C–O
  ||
  O
```

"-ane" becomes "-oic acid."

Ethanoic acid is commonly called acetic acid (in vinegar).

Start with butane C–C–C–C and you get

```
C–C–C–C–O
      ||
      O
```

butanoic acid, which is the smell of rancid butter.

The simplest carboxylic acid

```
C–O
||
O
```

is the acid in ant venom.

Since the title of this chapter is "More Naming," it's hard to resist pointing out that the carboxylic acids replace three H's with a carbonyl group (C=O) and a hydroxyl group (O–H). Those two groups together are called a **carboxyl group**.

carbonyl + hydroxyl = carboxyl

Hey! I, your reader, am going nuts. This is the tenth page of this chapter. Most chapters have been four pages plus two pages of Your Turn to Play.

Nuts? Me too. We've got three more classes of organic compounds which will add O's and N's, and we will be done.

It may help to think of it this way: In an organic chemistry class, this is all just introductory material. They then spend the semester showing the procedures to get from one class to another chemically.

You have the -anes (methane, ethane, propane), the branched -anes (like 3-ethylheptane), the -enes (which contain C=C), the -ynes (which contain C≡C), the ring compounds (like benzene), *and* those nine classes. Imagine memorizing the chemical pathways among all those things!

I don't want to imagine. This sounds as complicated as learning about all the different kinds of beetles in a biology class.*

I couldn't have expressed that better.

Okay. Let's get it done.

Class #7: The Ethers

C–O–C

The most famous is diethyl ether. In the old days they would give patients ether to knock them out before surgery. In the really old days, surgery was done without anesthetic. That was painful.

Class #8: The Esters

```
C–C–O–C
  ||
  O
```

the suffix is "oate."

Finally, something that smells decent.

```
C–C–O–C–C–C
  ||
  O
```

pears

```
C–C–O–C–C–C–C–C
  ||
  O
```

bananas

```
C–C–O–C–C–C–C–C–C–C–C
  ||
  O
```

oranges

```
C–C–C–C–O–C–C
      ||
      O
```

ethyl butanoate

pineapples

```
C–C–C–C–O–C–C–C–C–C
      ||
      O
```

apricots

* It's estimated that there are about 1,000,000 different species of beetles.

To "simplify" (ha-ha) things, if there is a C–C(=O)–O–C–C–C at one end, that circled part is called **acetate**.

Instead of propyl ethanoate, it can be named propyl acetate.

3 carbon (propyl) 2 carbon (ethanoate)

Class #9: The Amides

Look for C–C(=O)–N

If the N has C on only one end, then the naming is easy.

C–C(=O)–NH_2 methanamide

C–C–C(=O)–NH_2 ethanamide

If there are C's on both sides of the N, –C–C(=O)–N–C–

then we got *Life*! . . . and biochemistry.

Chapter Thirty-six
Biochemistry

Fred looked at the clock again. Good teachers do not go beyond the class hour. It was five minutes till eleven. That was plenty of time for him to "do" biochemistry. He knew he might leave out some minor points, but he would hit all the major points.

Well . . . some of the major points. This was understandable since some biochemists spend their entire lives studying one or two minor points.

The major points include proteins, carbohydrates (sugars, starches and cellulose), lipids (fats and oils), steroids, and nucleic acids (DNA and RNA).

"Protein," Fred began, "is the biggie. Dry out your body and what's left is 50% protein."

Joe took a big swallow of Sluice. He didn't want his body to "dry out."

"Making a protein is straightforward. You start with some amino acids* and you link them together with the amide group. [C–C–N]
(the middle C is double-bonded to O)

"The amino acids all have cute little nicknames. Glycine is call Gly. Alanine is called Ala. Leucine is called Leu."

One of the reporters in the back of the room began composing a country-western song, "I went huntin' for my Leu and found her in my protein soul." Next week it would hit the top of the charts and make millions for that song writer.

"Here is the start of the human hemoglobin protein Leu‿Ser‿Pro ‿Ala‿Asp‿Lys‿. . . . I'm using‿to indicate the amide group links. Some people just use hyphens. Chemists call those‿the **peptide bonds**.

"Proteins are chains of at least 50 amino acids."

Joe had unplugged his popcorn machine and plugged in his electric griddle. He was busy frying three hamburger patties. The buns were in his toaster.

---

* The amino acids are the building blocks for proteins. There are 22 of them in nature. We use 20 of them in our bodies.

"When you consume protein, say in the form of hamburgers, all the peptide bonds are broken and the cute little amino acids—Leu, Ser, Pro, Ala, . . .—enter your blood stream on the way to your cells.

"Your body then reassembles those amino acids into the proteins that it needs. Some proteins are used for structure: nails, hair, tendons. Some for movement: muscles. Some for transport: hemoglobin moves oxygen in the blood. Some for hormones. Some for fighting infections: antibodies.

"Your body makes over 100,000 different proteins. Be very glad that you don't have to consciously assemble each of them from the 20 amino acids."

Joe said, "That would take too long." He was right.

"If you had 20 buckets of beads (the amino acids) and you had to string together the protein necklaces, each of which is at least 50 beads long, you would have to be very careful. One mistake, one wrong bead, and you might have a radically different protein."

Joe said, "Picky, picky, picky," and he was right.

Ashley raised his hand. "I know I don't play the bead game consciously. I don't have to think about it. Who—or what—is doing the assembling of the amino acids into all those proteins?"

"James Watson and Francis Crick figured this out in 1953. The blueprints for constructing those zillions of different proteins that you need are the super giant DNA molecules. Each of them 'weighs' between six million and sixteen million amu. Recall 12 atomic mass units is the weight of a carbon atom.

"In the nucleus of each cell sits super giant DNA molecules. In another part of the cell is a ribosome which is the factory where the protein is made. There are three kinds of RNA molecules running around. The rRNA (ribosomal RNA) works in the ribosome factory putting together proteins. The mRNA (messenger RNA) first runs to the blueprint DNA to get a copy of the protein's formula and then to the ribosome factory. The tRNA (transfer RNA) runs around in the cell picking up amino acids and taking them to the factory." On the board Fred drew . . .

The pictures of cowboys Fred had cut out of a magazine. The other drawings were his.

"The big mistake in teaching chemistry or biology is to get lost in the details. Are you really learning something important—something that can fit into the rest of your life—or is it just details?

"Many chem textbooks splash out on a whole page the stick formulas for all the amino acid cows that will get herded to the ribosome factory. Why? Even learning their names doesn't do much.* What is much more important is that eleven of the amino acids your body can manufacture on its own, but nine of them must come from your diet. If you don't eat them, the tRNA cowboy would not be able to find them in the cell and herd them off to the ribosome protein factory. There will be proteins—essential proteins—that won't get made.

"Essential amino acids—Val and Leu and seven others—gotta be in your diet or your nails and hair will look ugly [important for Darlene], your muscles will get weak [important for Ashley], and your antibody production will cease [important only for those who want to avoid fatal infections].

"Some foods are called **complete proteins** because they contain all nine essential proteins in the ① correct quantity and in the ② correct ratio needed by humans. Examples are meats, fish, poultry, dairy, and eggs. Incomplete proteins are found in grains, beans, legumes, seeds, nuts, and corn. Corn, for example, is very low in Lys and Ile.

"A diet of incomplete proteins can provide all the nine essential amino acids *if* you make the right food combinations. For example, the traditional Cajun dish of red beans and rice will work."

It never seems to have occurred to six-year-old Fred that the lack of complete protein in his diet might be the reason he is only 36 inches tall.

It has not occurred to Joe that his diet will result in many years of infirmity and death in his fifties.

Fred looked at the clock. He had one minute left. He could talk about optimal human diets—high carbs vs. low carbs. He could talk about saturated fats (all single bonds between carbon atoms).

---

* Glycine (Gly), Alanine (Ala), Valine (Val), Isoleucine (Ile), Leucine (Leu), Methionine (Met), Phenylalanine (Phe), Proline (Pro), Tryptophan (Trp), Serine (Ser), Threonine (Thr), Tyrosine (Tyr), Cysteine (Cys), Asparagine (Asn), Glutamine (Gln), Aspartic acid (Asp), Glutamic acid (Glu), Lysine (Lys), Histidine (His), Arginine (Arg).

He loved the very complex field of protein folding. After the ribosome makes a protein, the protein scrunches up into a wrinkly mess that makes a plate of spaghetti seem tame.

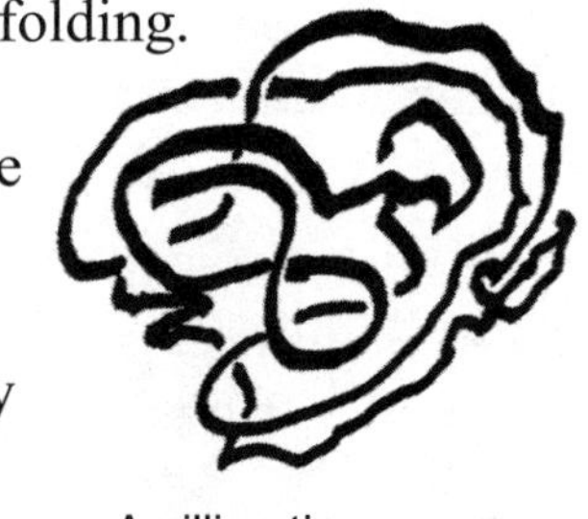

A zillion times more complicated than this.

Given the **primary structure of a protein** (its sequence of amino acids), it is mind-bogglingly difficult to predict is secondary, tertiary and quaternary structure—in other words, how it will fold up.

How it folds up determines what the protein will do.

If it misfolds (←a technical word), it can become inactive or even toxic.

Suddenly, the room went **dark**. No one knew why. The class hour was over. The students left carefully, not wanting to trip over the extension cords— especially the one that went to electric pizza oven that Joe had just plugged in.

Index

Index

Index

Index